Felix Hoos

Compact and efficient femtosecond supercontinuum sources

Felix Hoos

Compact and efficient femtosecond supercontinuum sources

Compact and efficient femtosecond supercontinuum sources based on diode-pumped solid-state lasers

Südwestdeutscher Verlag für Hochschulschriften

Impressum/Imprint (nur für Deutschland/ only for Germany)
Bibliografische Information der Deutschen Nationalbibliothek: Die Deutsche Nationalbibliothek verzeichnet diese Publikation in der Deutschen Nationalbibliografie; detaillierte bibliografische Daten sind im Internet über http://dnb.d-nb.de abrufbar.

Alle in diesem Buch genannten Marken und Produktnamen unterliegen warenzeichen-, marken- oder patentrechtlichem Schutz bzw. sind Warenzeichen oder eingetragene Warenzeichen der jeweiligen Inhaber. Die Wiedergabe von Marken, Produktnamen, Gebrauchsnamen, Handelsnamen, Warenbezeichnungen u.s.w. in diesem Werk berechtigt auch ohne besondere Kennzeichnung nicht zu der Annahme, dass solche Namen im Sinne der Warenzeichen- und Markenschutzgesetzgebung als frei zu betrachten wären und daher von jedermann benutzt werden dürften.

Verlag: Südwestdeutscher Verlag für Hochschulschriften Aktiengesellschaft & Co. KG
Dudweiler Landstr. 99, 66123 Saarbrücken, Deutschland
Telefon +49 681 37 20 271-1, Telefax +49 681 37 20 271-0, Email: info@svh-verlag.de
Zugl.: Stuttgart, Universität Stuttgart, Diss., 2009

Herstellung in Deutschland:
Schaltungsdienst Lange o.H.G., Berlin
Books on Demand GmbH, Norderstedt
Reha GmbH, Saarbrücken
Amazon Distribution GmbH, Leipzig
ISBN: 978-3-8381-1073-8

Imprint (only for USA, GB)
Bibliographic information published by the Deutsche Nationalbibliothek: The Deutsche Nationalbibliothek lists this publication in the Deutsche Nationalbibliografie; detailed bibliographic data are available in the Internet at http://dnb.d-nb.de.

Any brand names and product names mentioned in this book are subject to trademark, brand or patent protection and are trademarks or registered trademarks of their respective holders. The use of brand names, product names, common names, trade names, product descriptions etc. even without a particular marking in this works is in no way to be construed to mean that such names may be regarded as unrestricted in respect of trademark and brand protection legislation and could thus be used by anyone.

Publisher:
Südwestdeutscher Verlag für Hochschulschriften Aktiengesellschaft & Co. KG
Dudweiler Landstr. 99, 66123 Saarbrücken, Germany
Phone +49 681 37 20 271-1, Fax +49 681 37 20 271-0, Email: info@svh-verlag.de

Copyright © 2009 by the author and Südwestdeutscher Verlag für Hochschulschriften Aktiengesellschaft & Co. KG and licensors
All rights reserved. Saarbrücken 2009

Printed in the U.S.A.
Printed in the U.K. by (see last page)
ISBN: 978-3-8381-1073-8

Compact and efficient femtosecond supercontinuum sources based on diode-pumped solid-state lasers

Von der Fakultät für Mathematik und Physik
der Universität Stuttgart
zur Erlangung der Würde eines Doktors der
Naturwissenschaften (Dr. rer. nat.) genehmigte Abhandlung

vorgelegt von
Felix Hoos
aus Heilbronn

Hauptberichter: Prof. Dr. Harald Giessen
Mitberichter: Prof. Dr. Peter Michler
Tag der mündlichen Prüfung: 24. April 2009

Physikalisches Institut der Universität Stuttgart
2009

Zusammenfassung

Diese Arbeit beschäftigt sich mit der Realisierung von kompakten und multi-Watt Femtosekunden-Superkontinuumquellen, welche auf Kombinationen von diodengepumpten Laseroszillatoren und nichtlinearen Glasfasern basieren. Der Schwerpunkt der Arbeit liegt auf der Entwicklung von direkt diodengepumpten sub-200 fs Laseroszillatoren, die sich als Pumplaser für solche Fasern eignen. Der weitere Teil beschreibt eine umfassende Studie der Eigenschaften von Superkontinua, welche bei Pumpwellenlängen um 1 μm erzeugt werden.

Mikrostrukturierte Fasern, entweder gezogene Fasern ("tapered fibers") oder photonische Kristallfasern (PCF), können schmalbandiges Laserlicht zu teilweise mehreren Oktaven breiten Spektren konvertieren. Dabei können Effizienzen von über 50% und Durchschnittsleistungen von mehreren Watt erreicht werden. Die Eigenschaften solcher Spektren hängen stark von der Länge der Eingangspulse ab. Ein besonderes Merkmal im Femtosekundenregime, welches durch eine Obergrenze von 200 fs für die Eingangspulsbreite festgelegt werden kann, ist eine hohe spektrale Phasenkohärenz.

Der Hauptteil dieser Arbeit beschreibt die Entwicklung von zwei diodengepumpten sub-200 fs Laseroszillatoren. Einerseits die eines sehr kompakten 20 MHz Yb:Glas Lasers, andererseits die eines multi-Watt 44 MHz Yb:KGW Oszillators. Mit dem letzteren Oszillator konnten wir, nach unseren Kenntnissen, die bisher höchste optisch-zu-optische Effizienz eines diodengepumpten multi-Watt Lasers mit einer Pulsdauer von 150-170 fs erzielen. Diese Fortschritte beruhen auf den Ergebnissen unserer Forschung an Yb:KGW Streifenlasern und dem Einsatz von Breitstreifendiodenlasern. Hierzu führten wir detaillierte experimentelle sowie numerische Untersuchungen der thermischen Effekte in Yb:KGW Kristallstreifen durch, die durch Pumpen mit diesen Dioden entstehen. Im zweiten Teil dieser Arbeit wurden die Eigenschaften von Superkontinua untersucht, die in gezogenen Fasern mit für Ytterbium-Lasern typischen Wellenlängen um 1 μm erzeugt wurden. Während für das Intensitätsrauschen und den Einfluss der Fasergeometrie auf die Spektren vergleichbare Ergebnisse wie in früheren Arbeiten mit Ti:Saphir Lasern gefunden wurden, ließ sich bei Ytterbium-

Lasern ein Unterschied beim Einfluss des Pulschirps auf die Spektren feststellen.

Abstract

This thesis deals with the realization of compact and multi-Watt femtosecond supercontinuum sources based on diode-pumped laser oscillators and highly nonlinear glass fibers. The main aspect of this work focuses on the development of directly diode-pumped sub-200 fs laser oscillators which are suitable as pump lasers for those fibers. In addition, a comprehensive study of the properties of supercontinua generated with pump wavelengths around 1 μm is presented.

Micro-structured fibers, either tapered fibers or photonic crystal fibers (PCF), can convert narrow band laser light into multiple octave broad spectra. Average powers of several Watts and conversion efficiencies of more than 50% can be achieved. The properties of those spectra depend strongly on the input pulse duration. A particular feature of the femtosecond regime, which can be defined by an input pulse duration shorter than 200 fs, is a high spectral phase coherence.

The main part part of this thesis describes the development of two diode-pumped sub-200 fs laser oscillators. On one hand, a very compact 20 MHz Yb:glass laser, on the other hand, a multi-Watt 44 MHz Yb:KGW oscillator. With the latter oscillator, we demonstrated to our knowledge the highest optical-to-optical efficiency of a diode-pumped multi-Watt laser with a pulse duration of 150-170 fs. These advances are based on our study of Yb:KGW slab-oscillators pumped by broad-area laser diodes. The thermal effects in the Yb:KGW crystal slabs pumped by high-power broad-area diodes were investigated in detail experimentally as well as numerically. In the second part of this work, we studied the properties of supercontinua generated in tapered fibers with Ytterbium lasers with typical wavelengths around 1 μm. While the intensity noise and the influence of the fiber geometry on the spectra were found to be similar to former results achieved with Ti:sapphire lasers, a different influence of a pulse pre-chirp on the spectra was observed for Ytterbium lasers.

Publications

Part of this work have already been published:

In scientific journals:

- F. Hoos, T.P. Meyrath, S. Li, B. Braun, and H. Giessen, *Femtosecond 5 W Yb:KGW slab laser oscillator pumped by a single broad-area diode and its application as supercontinuum source*, Appl. Phys. B (in press), DOI: 10.1007/s00340-009-3421-3.

- F. Hoos, S. Li, T.P. Meyrath, B. Braun, and H. Giessen, *Thermal lensing in an end-pumped Yb:KGW slab laser with high power single emitter diodes*, Opt. Express **16**, 6041 (2008).

- F. Hoos, S. Pricking, and H. Giessen, *Compact portable 20 MHz solid-state femtosecond whitelight-laser*, Opt. Express **14**, 10913 (2006).

At scientific conferences:

- F. Hoos, T.P. Meyrath, S. Li, B. Braun, and H. Giessen, *Efficient and Simple High-Power Femtosecond Yb:KGW Slab-Laser Pumped By A Single Broad-Area Diode*, Advanced Solid State Photonics (ASSP 2009), MB13, Denver, CO, USA (2009).

- F. Hoos, T.P. Meyrath, S. Li, B. Braun, and H. Giessen, *Highly efficient 4 W, 160 fs Yb:KGW slab-laser oscillator pumped by a single broad area and its application for supercontinuum generation*, Annual Meeting of the Deutsche Physikalische Gesellschaft (DPG), Q 4.6, Hamburg, Germany (2009).

- F. Hoos, B. Braun, and H. Giessen, *Towards a compact high-average-power femtosecond supercontinuum source*, Annual Meeting of the Deutsche Physikalische Gesellschaft (DPG), Q 11.2, Darmstadt, Germany (2008).

Contents

Zusammenfassung   i

Abstract   iii

Publications   v

Contents   vii

1. Introduction   1

2. Propagation of ultra-short laser pulses   5
    2.1. Group velocity (GVD) and group delay dispersion (GDD)   6
    2.2. Self-phase modulation (SPM)   7
    2.3. Optical solitons   8

3. Generation of femtosecond supercontinua   11

4. Solitary mode-locked laser oscillators   17
    4.1. Dynamics of mode-locked laser oscillators   18
        4.1.1. Laser rate equations and master equation of mode locking . .   18
        4.1.2. Suppression of Q-switching   20
        4.1.3. Initiation and self-starting of mode-locking with saturable absorbers   20
    4.2. Solitary mode-locking with semiconductor saturable absorber mirrors (SESAMs)   21
    4.3. Pulse break-up and the B-integral   23

5. Resonator design for diode-pumped femtosecond slab lasers   25
    5.1. Numerical model of the thermal effects in longitudinally diode-pumped gain media   25

	5.1.1.	Retrieval of the focal length of the thermal lens	29
5.2.	Simulation of the resonator internal beam waist	30

6. Compact 20 MHz femtosecond supercontinuum system based on an Yb:glass oscillator — 35

6.1.	Current state of research	35
6.2.	Properties of Yb:glass as laser medium	36
6.3.	Experimental setup of the Yb:glass oscillator	39
	6.3.1. Laser medium and pump configuration	39
	6.3.2. Layout of the laser resonator	40
	6.3.3. Design of the Herriott-type multi-pass cell	43
	6.3.4. Solitary mode-locking and SESAM	45
6.4.	Characterization of the laser parameters	46
6.5.	Supercontinuum generation with the Yb:glass laser	49
	6.5.1. Measurements with tapered fibers	49
	6.5.2. Noise measurements	52
	6.5.3. Influence of pulse chirp and fiber pig-tails	53
	6.5.4. Measurements with photonic crystal fiber (PCF)	55
6.6.	Conclusion	57

7. Multi-Watt 44 MHz femtosecond supercontinuum source based on an Yb:KGW oscillator — 59

7.1.	Current state of research on multi-Watt sub-200 fs laser oscillators	60
7.2.	Ytterbium doped double tungstates	61
7.3.	Femtosecond 5 W Yb:KGW slab oscillator	62
	7.3.1. Properties of Yb:KGW	62
	7.3.2. Broad-area laser diodes	65
	7.3.3. Geometry, orientation and properties of the Yb:KGW slab	70
	7.3.4. Experimental characterization of the thermal effects and cw-operation of the Yb:KGW oscillator	71
	7.3.5. Numerical modeling of the thermal lens induced by pumping with broad-area diodes	77
	7.3.6. Comparison between N_g-cut and athermal orientation of Yb:KGW for the slab-laser geometry	86
	7.3.7. Experimental setup of the femtosecond 44 MHz oscillator	88
	7.3.8. Mode-locked properties of the Yb:KGW oscillator	91
7.4.	Generation of high-power femtosecond supercontinua	95

	7.4.1. Supercontinua generated by tapered fibers	95
	7.4.2. Supercontinuum generation in polarization-maintaining PCF	97
7.5.	Conclusion	99

8. Summary   101

A. List of acronyms and abbreviations   105

B. List of symbols and constants   107

C. List of ray-matrices used for resonator calculations and thermal lens retrieval   111

Bibliography   113

Acknowledgements   125

Chapter 1
Introduction

In 2000, Ranka et al. discovered the possibility of generating multiple octave broad continuous spectra (or more precisely, broad and dense frequency combs) in air-silica micro-structured fibers with nanojoule femtosecond laser pulses. The key point was to provide anomalous dispersion around the pump wavelength which could be done by tailoring the fiber geometry [1]. Such continuous and broad spectra, so called supercontinua (SC), were known before and could be generated for example by focusing high-energy laser pulses onto thin sapphire plates. Micro-structured fibers have shown the possibility of increasing the average power of such spectra tremendously. Only few months after the publication of Ranka et al., Birks and his co-workers realized supercontinuum generation by pumping a 90 mm long tapered fiber with nanojoule femtosecond laser pulses [2]. The involved optical processes were similar in this case. Anomalous dispersion was achieved by drawing a piece of common glass fiber down to diameters of a few microns. The wave-guiding in such thin fibers is then due to reflections off the glass-air interface which changes the fiber dispersion. While so-called photonic crystal fibers (PCF) used by Ranka et al. can be manufactured very homogeneously over nearly arbitrary lengths, tapered fibers have the advantage that the manufacturing does not require any complex or expensive equipment.

Supercontinuum sources soon proved to be a versatile tool for numerous applications ranging from the use as broad tunable laser source for spectroscopy or microscopy to the generation of broadband frequency combs. Thus the research on both kinds of micro-structured fibers attracted much attention. The physical background and the complex dynamics of supercontinuum generation in micro-structured fibers was investigated theoretically [3, 4, 5] and experimentally, see for example refs. [6, 7]. Today, the generation process of supercontinua in micro-structured fibers is well understood. The dynamics can be described by a model based on a generalized nonlinear Schrödinger equation which is valid for different regimes of laser pulse durations [3]. Differences between these regimes can be predicted. In addition to supporting the theoretical background, many experiments also focused on the realization of different supercontinuum sources, for instance with many different pump wavelengths [8, 6, 9, 10, 11], various pump pulse durations from nano- to femtoseconds [6] or even with cw-light [12]. Furthermore, several approaches were performed in order to increase the average power. Recent publications report up to several tens of Watts of pump and output power [13, 14, 15].

In some cases, the technology of supercontinuum sources has reached a quite mature and even industrially suitable level. For example, microchip laser based nanosecond sources or fiber laser based picosecond sources capable of generating average powers of up to 8 W are commercially available. However, there is still a lack of high-power femtosecond systems due to the challenging task of providing a convenient pump source which generates high-power laser pulses shorter than 200 fs. Nonetheless, such femtosecond sources would provide some properties that are beneficial for various experiments. One major difference to longer pulse regimes is that the grade of spectral coherence between two successive laser pulses and the stability of the phase relation between two different spectral components are tremendously higher for short femtosecond pump pulses [3, 16]. Both factors greatly increase for pulses shorter than 200 fs. Example applications which would benefit from these properties are experiments aiming at temporally compressing such multi-octave broad spectra or interferometric methods, for instance heterodyne coherent anti-Stokes Raman spectroscopy (CARS) imaging [17]. In addition to the aforementioned supercontinuum properties, the use of femtosecond pulses also offers higher peak powers and a higher temporal resolution than picosecond sources. This is useful for multi-photon microscopy [18] or temporally resolved measurements, such as pump and probe techniques for example [19].

Up to now, femtosecond supercontinuum sources have been restricted to physics laboratories, could usually provide average powers of not more than a few hundred milliwatts, and have required complex pump lasers. The aim of this work was to provide concepts and solutions for compact, cost effective, and easy to use femtosecond supercontinuum sources and to increase the average power of such sources. The work presented here is split between two different areas of research. One part of this thesis deals with the development of diode-pumped sub-200 fs laser oscillators which are needed as pump sources. In the second part of this work, the properties of supercontinua that can be obtained with the developed pump lasers were investigated. The experimental results of both parts are presented in Chapters 6 and 7 of which each presents the development of a complete femtosecond supercontinuum source. Chapter 6 describes an approach for realizing a very compact supercontinuum system by employing an Yb:glass laser and folding the oscillator footprint by employing a Herriott-type multi-pass cell. In Chapter 7 we present experimental results of a diode-pumped 5 W sub-200 fs laser oscillator. With this oscillator, femtosecond supercontinua with average powers of up to 1.3 W can be obtained with tapered fibers. Both realized supercontinuum systems are based on pumping micro-structured fibers with Ytterbium lasers. The results of Chapters 6 and 7 also provide a comprehensive study of properties relevant for practical applications such as noise characteristics, polarization stability, and the influence of a pulse pre-compression for pump wavelengths of Ytterbium lasers around 1 μm.

A major topic of this thesis is the study of diode-pumped high-power sub-200 fs laser sources that can be utilized with micro-structured fibers. Thin-disk lasers would be well suited for this purpose, however, the pulse duration with this class of lasers seems to be limited to durations longer than 200 fs [20]. Another promising concept is the class of Ytterbium doped tungstate slab lasers [21, 22, 23]. In Chapter 7 we present the results of our research on this class of lasers and show that progress in simplicity and efficiency is possible by the use of recently available high-power broad-area diodes [24, 25, 26] as well as by employing the N_g-cut orientation of Yb:KGW. An important issue in the realization of this class of lasers are the influences of thermal effects. We experimentally and numerically investigated the thermal effects induced by broad-area diodes in a thin Yb:KGW crystal slab and could find a crystal orientation for which these effects can be very well compensated and a nearly diffraction limited laser beam can be obtained for pump powers of up to 18 W. A first realization is shown by the 5 W Yb:KGW oscillator. In Chapter 5, the necessary numerical tools for the resonator design and for simulating the expected thermal effects are presented.

Chapter 2
Propagation of ultra-short laser pulses

The physics of the propagation of ultra-short laser pulses is of great importance for both the design of femtosecond lasers as well as the description of supercontinuum generation in micro-structured fibers.

Depending on the position along the beam waist, continuous-wave (cw) Gaussian laser beams can approximately be considered either as a plane or as a spherical electro-magnetic wave. Nonetheless, in both cases the propagation of a cw laser beam is sufficiently well described in terms of linear optics. In contrast to that, laser pulses need to be regarded as wave packets and nonlinear optical effects need to be taken into account. In the context of the slowly varying field envelope approximation (SVEA) [27], the electric field $E(\mathbf{r},T)$ of a laser pulsed propagating along the z-direction can be expressed by

$$E(\mathbf{r},T) = A(\mathbf{r},T) \cdot e^{-i(\omega_0 \cdot T - k(\omega_0) \cdot z)} + c.c., \qquad (2.1)$$

with the complex field amplitude $A(\mathbf{r},T)$, the global time T, the spatial coordinate $\mathbf{r}$, the carrier angular frequency ω_0, and the wave vector $k(\omega_0)$. Especially in regard to mode-locking, the field amplitude $A(\mathbf{r},T)$ can be given in the time domain as a superposition of longitudinal laser modes as

$$A(\mathbf{r},T) = \sum_m E_m(\mathbf{r},T) \cdot e^{-i((\omega_m - \omega_0)T - (k(\omega_m) - k(\omega_0))z + \Phi_m)}, \qquad (2.2)$$

where E_m are the field amplitudes of the individual modes, ω_m and $k(\omega_m)$ are the angular frequencies and wave vectors of each mode, respectively, and Φ_m are phase offsets. Often the transversal spatial field distribution and the temporal envelope can be separated. In this case, only the propagation along the z-axis is included and the complex amplitude is given by $A(z,T)$ which is related to $A(\mathbf{r},T)$ by $A(\mathbf{r},T) = F(x,y) \cdot A(z,t)$, where $F(x,y)$ describes a time-independent lateral intensity distribution.

Furthermore, the description of the field envelope can be simplified by expressing Eq.2.2 in terms of a retarded time t as $A(t,T)$ or as $A(z,t)$. The parameter t describes a co-moving time frame with t being a retarded time with the zero position always at the pulse center. It is related to the global time T by $t = T - z/v_g$, where v_g is the group velocity. The amplitudes are normalized in such a way that the instantaneous power can be derived by $|A(z,t)|^2 = P(z,t)$. In the description of Eq. (2.1), all effects that change the shape of the field amplitude or the optical spectrum can be described by their influence on the complex term $A(z,t)$.

Such wave-packets possess two interesting properties that concern the propagation in media, namely extremely high peak intensities and a broad optical spectrum. The high intensities give rise to the excitation of nonlinear effects, whereas the broad optical spectrum leads to dispersion of the wave-packets due to the wavelength-dependence of the refractive index. The most-prominent nonlinear effect concerning femtosecond lasers is the optical Kerr-effect. The Kerr-effect causes two phenomena, self-phase modulation (SPM) and the formation of a Kerr-lens. Of course, also other nonlinear effects can be excited. This is of particular importance for supercontinuum generation, as discussed in Chapter 3. The following section deals with the effect of SPM and dispersion of ultra-short laser pulses. The formation of a Kerr-lens is treated in the section about resonator design for femtosecond lasers (Chapter 5.2).

2.1. Group velocity (GVD) and group delay dispersion (GDD)

As discussed above, the electric field of a laser pulse can be described by a product of a slowly-varying envelope $A(z,t)$ and field oscillations with a frequency ω_0 (Eq.(2.1)). However, the index of refraction of a medium is usually dependent on the frequency, i.e. $n = n(\omega)$. Thus, the wave-vector $k(\omega) = n(\omega) \cdot k_0$ is also wavelength dependent and varies for different spectral components ω_m. This fact needs to be taken into account for the temporal behavior of the envelope $A(z,t)$. For example, if in a certain medium red light propagates faster than blue light, the longer wavelength components of a laser pulse will lead the pulse while the blue components will trail behind. Such a laser pulse broadens temporally during the propagation in such a medium. A mathematical description can be found by assuming the optical bandwidth $\Delta\omega$ of a laser pulse to be small compared to the center frequency ω_0. In this case, term with $k(\omega_m)$ in Eq. 2.2 can be expressed by extending the wave-vector

2.2. Self-phase modulation (SPM)

$k(\omega)$ in a Taylor-series around the center frequency ω_0

$$\begin{aligned}k(\omega) &= k(\omega_0) + \left.\frac{\partial k}{\partial \omega}\right|_{\omega_0} \cdot (\omega - \omega_0) + \frac{1}{2}\left.\frac{\partial^2 k}{\partial \omega^2}\right|_{\omega_0} \cdot (\omega - \omega_0)^2 + \frac{1}{6}\left.\frac{\partial^3 k}{\partial \omega^3}\right|_{\omega_0} \cdot (\omega - \omega_0)^3 + \cdots \\ &= k(\omega_0) + k_1 \cdot (\omega - \omega_0) + k_2 \cdot (\omega - \omega_0)^2 + k_3 \cdot (\omega - \omega_0)^3 + \cdots,\end{aligned}$$
(2.3)

with the length related material dispersion coefficients k_i. The first three coefficients k_1, k_2, and k_3 are referred to as group delay (GD), group velocity dispersion (GVD), and third-order dispersion (TOD). In order to describe the GVD of a whole bulk element, one often uses the so-called group delay dispersion (GDD) which is the product of the GVD and the propagation length in this medium.

To calculate the influence of the dispersion on the field envelope $A(z,t)$, one can regard Eq. (2.2) together with Eq. 2.3 in the Fourier space and derive a differential equation. Considering only dispersion up to the second order k_2, the influence of GVD on the pulse envelope can be expressed by the following differential equation [28]:

$$\frac{\partial A(z,t)}{\partial z} = i \cdot \frac{k_2}{2} \cdot \frac{\partial^2 A(z,t)}{\partial t^2}.$$
(2.4)

Eq. (2.4) can be solved analytically for certain envelopes, e.g. for Gaussian or sech-shaped envelopes. For more complex envelopes or numerical simulations, a numerical solution in the Fourier space is very convenient. Analytical solutions of Eq. (2.4) show that the principal shape of a Gaussian or a sech-shaped pulse is preserved. However, during propagation the pulse will broaden in time while the field oscillations experience a frequency chirp.

2.2. Self-phase modulation (SPM)

Self-phase modulation is a third-order nonlinear optical effect that is due to the intensity dependence of the index of refraction. This third-order effect can be considered as a certain kind of degenerate four-wave mixing which explains why new frequency components can be generated [27]. A more intuitive description of SPM can be given directly by regarding the intensity dependence of the refractive index. This dependence is described by the nonlinear index n_2 by the relation $n(I) = n_0 + n_2 \cdot I$. Laser pulses experience a different index of refraction for different positions along their temporal envelope. This is also the case for different lateral positions in the beam profile. However, both effects can be separated. The influence of the temporal envelope leads to SPM, while the influence of the spatial beam profile induces the formation of a Kerr-lens. Regarding only the temporal pulse envelope, the tails of a laser pulse will see a different index of refraction than the center of the pulse, thus the laser pulse will accumulate a different phase along its envelope. Since the frequency can be considered as the derivative of the phase, this effect can be interpreted as a change of the carrier frequency along the pulse envelope. Thus, SPM does not influence the temporal shape of the pulse envelope, but changes the optical spectrum by generating new frequency components. As in the description of the influence of dispersion, the influence of SPM on the complex term $A(z,T)$ can also be described by a differential equation [28]:

$$\frac{\partial A(z,t)}{\partial z} = i \cdot \gamma \cdot \left|A(z,t)\right|^2 A(z,t), \tag{2.5}$$

with the SPM coefficient $\gamma = \frac{n_2(\omega_0)\omega_0}{cA_{eff}}$ and the effective mode-area in the medium A_{eff}. The effect of SPM is the broadening and modulation of the optical spectrum. However, the temporal (intensity) pulse envelope remains unchanged.

2.3. Optical solitons

For femtosecond lasers or pulse propagation in optical fibers, it would be beneficial to suppress all distorting nonlinear effects and dispersion. Under certain circumstances, the second order dispersion GVD can be compensated by simultaneously occurring SPM. Usually, in the case of normal dispersion and with the photon energy far away from the energy band-gap, both effects lead to a pulse chirp in the same way, with a red-shifted leading tail and a blue-shifted trailing tail. However, in the presence of anomalous dispersion with $k_2 < 0$, both effects can cancel one another. The dynamics of the pulse envelope $A(z,t)$ in the presence of GVD and SPM can be described by the following differential equation [28]:

$$i \cdot \frac{\partial A(z,t)}{\partial z} = -\frac{k_2}{2}\frac{\partial^2 A}{\partial t^2} + \gamma |A|^2 A. \tag{2.6}$$

Due to the technical similarity of Eq. (2.6) to the Schrödinger equation in quantum mechanics, this equation is called nonlinear Schrödinger equation (NLSE). One can find analytical solutions to this equation, which are so-called optical solitons. In general one can find solutions of different order N. The fundamental solution with $N = 1$ is a pulse with a time-invariant sech-shaped field envelope [28]:

$$A(z,t) = A_0 \cdot \mathrm{sech}\left(\frac{t}{\tau}\right) e^{-i\gamma A_0^2 z}. \tag{2.7}$$

2.3. Optical solitons

Here, A_0 is the normalized field amplitude, γ is the SPM coefficient, and τ determines the duration of the pulse. The parameter τ is related to the FWHM pulse width by $\tau_{FWHM} = 1.76 \cdot \tau$. In addition to the fundamental soliton, there also exist solutions to Eq. (2.6) with an order of $N > 1$. Such solutions are called higher order solitons and play an important role in the supercontinuum generation with femtosecond pulses. These higher order solitons do not posses a time-invariant envelope, but repeat the shape of their envelope periodically after a certain propagation distance. In order to represent a solution to Eq. (2.6), the energy of a laser pulse $E_P = 2 \cdot A_0 \cdot \tau$ must be of a certain value. In general for a soliton of the order N, the following condition needs to be fulfilled [28]:

$$N^2 = \frac{\gamma A_0^2 \tau^2}{|k_2|}, N \in \mathbb{N}. \tag{2.8}$$

If the pulse energy cannot fulfill Eq. (2.8) exactly, a soliton of the next possible order will form, while the remaining surplus energy will form a dispersive pulse which broadens temporally during propagation. If the dispersive medium is long, for example in a glass fiber, the dispersive fraction will finally fade out and only a soliton of a certain number N remains.

Eq. (2.6) and its solutions were derived for continuous conditions with SPM and GVD appearing simultaneously. This situation is well fulfilled in glass fibers for example. However, in laser resonators the different effects will appear in discrete, lumped elements. For instance, SPM in the gain medium and GDD due to reflections off of the dispersive mirrors. One can show that in this case similar solutions exist as for the continuous case. The fundamental soliton will also have a sech-shaped field envelope (i.e. a sech^2-shaped intensity envelope). However, the envelope is not invariant at every position in the laser resonator. Therefore, it is possible that laser pulses that are emitted from a laser oscillator are not transform-limited but slightly chirped.

Chapter 3
Generation of femtosecond supercontinua

A very convenient and stable method to generate supercontinua with high average powers can be realized by propagating narrow line-width laser light through microstructured fibers. The fiber could either be a photonic crystal fiber (PCF) [1, 29] or a tapered fiber [2, 8, 7]. In the femtosecond regime, propagation lengths of only a few cm in these fibers are necessary to convert a narrow laser spectrum to a multiple octave broad supercontinuum [3]. Since the physical effects causing this spectral broadening are similar in PCF and tapered fibers, the following discussion focuses on femtosecond supercontinuum generation in tapered fibers.

Tapered fibers can be drawn from short pieces of regular SMF28 glass fiber in a home-built drawing machine. The structure of a tapered fiber, schematically shown in Fig. 3.1, consists of three parts: two pig-tails of regular SMF28 fiber, the taper region, and the thin waist region. The diameter of the waist region is only a few

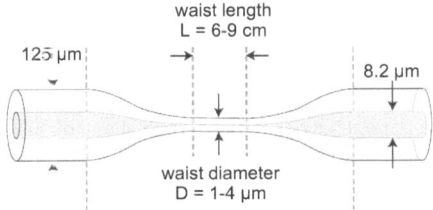

Figure 3.1.: Schematic structure of a tapered fiber. The thin waist region with diameters in the range between 1-4 μm and lengths around 6-9 cm is connected with SMF28 fiber pig-tails by the tapered regions.

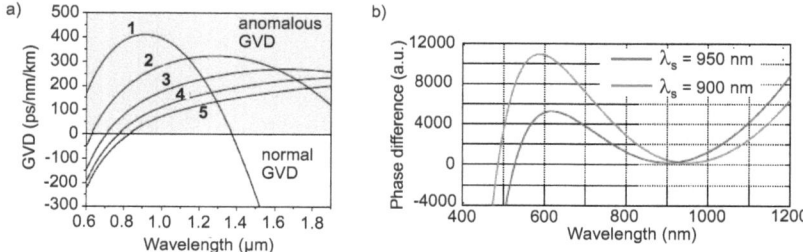

Figure 3.2.: a) Wavelength dependence of the GVD in tapered SMF28 fibers. The waist region is surrounded by air and the waist diameters are (1) 1 μm, (2) 1.5 μm, (3) 2 μm, (4) 2.5 μm and (5) 3 μm [30]. b) Calculation of the phase difference between a fundamental soliton at the wavelength λ_S and the dispersive wave for a tapered fiber with a diameter of 2.2 μm and an input power of 100 mW [31].

μm and ranges between 1-5 μm for practical applications. In the waist region, the light is no longer guided by the core-cladding interface, but the core is negligibly thin thus that the light is guided by the glass-air interface. The length of the waist region is restricted by the manufacturing process. With our manufacturing setup, waist lengths of up to 30 cm can be achieved. However, in the femtosecond regime, waist lengths of 6-9 cm are sufficient. During the manufacturing process, the glass of the fiber is softened by heating it with a burner. Then the fiber is drawn apart and elongated slowly at a constant speed. After a certain elongation, the burner is simultaneously moved in such a way that the heating zone is enlarged while the fiber is further elongated. The shape of the taper transitions as well as the waist diameter and length can be reproducibly adjusted by controlling the movement of the burner [33]. In this work, we chose a linearly increasing heating zone resulting in taper transitions with a diameter that is proportional to $D(z) \propto D_0(1 + \frac{2}{L_0}\frac{\alpha}{1-\alpha} \cdot z)^{-1/2\alpha}$, where L_0 is the initial width of the heating zone, e.g. the width of a flame. The parameter α determines the shape of the taper transition. It can be chosen arbitrarily between $-1 \cdots +1$ [33]. In this work, we chose $\alpha = 0.8$ for all manufactured tapered fibers. Over the waist region, an approximately constant diameter is achieved.

Tapered fibers possess two properties that cause the broadening of narrow linewidth laser light into supercontinua. On one hand, due to the extremely small waist diameters, the light is concentrated onto a small mode area over a long propagation

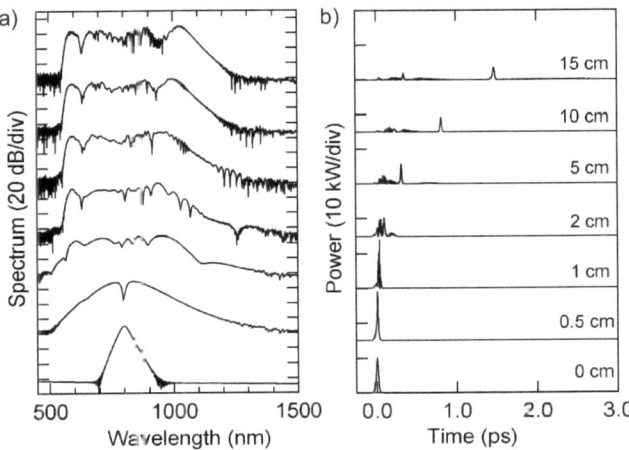

Figure 3.3.: a) Simulation of the temporal dynamics of the optical spectrum. b) Simulation of the dynamics of the pulse envelope. Both calculations were performed for a laser pulse with a duration of 50 fs for the propagation over a distance of 15 cm in a PCF [32].

length in a nonlinear material. This supports the excitation of nonlinear optical processes. On the other hand, and of greater importance, the GVD of these fibers can be tailored by tapering them, changing the diameter, and guiding the light by a glass-air interface. Fig. 3.2 (a) shows the wavelength-dependent GVD of tapered fibers with different waist diameters. The GVD in this diagram is not given in the unit fs^2/m as it is common in the field of ultra-fast lasers, but in the unit ps/km/nm as it is widely used in telecommunication applications. Both descriptions can be transferred into each other by the relation $k_2(fs^2/m) = -\frac{\lambda^2}{2\pi c} \cdot k_2(ps/nm/km)$. For supercontinuum generation it is essential that light at the pump wavelength experiences anomalous dispersion. Solitons that form at common intensities are usually of orders $N > 1$. The dynamics of these higher order solitons play an important role for supercontinuum generation.

An understanding of the physical processes of supercontinuum generation, and the quite complex interaction between them, was gained by experiments as well as numerical simulations of supercontinuum generation [3, 6, 4]. A, therefore, necessary mathematical description of pulse propagation in tapered fibers and PCF was found

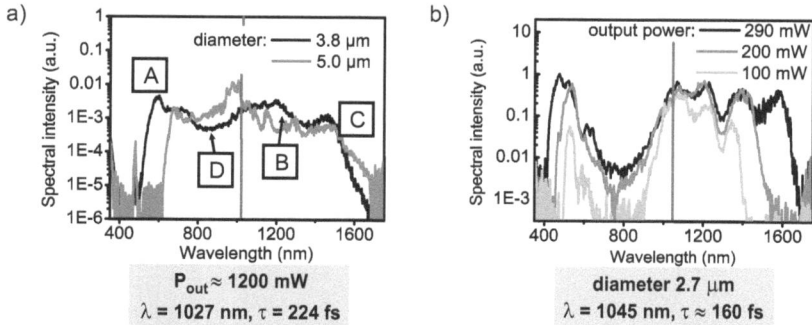

Figure 3.4.: Measured spectra of supercontinua generated for different conditions: a) At a constant output power of approximately 1.2 W, but for two different fibers with waist diameters of 3.8 μm (black line) and 5 μm (gray line), respectively. b) Supercontinua generated in the same tapered fiber with 2.7 μm waist diameter, but for different output powers. The red lines indicate the position of the pump wavelength. (A: peak of the dispersive wave, B and C: peaks of red-shifted solitons, D: spectral gap, filled by four-wave mixing, e.g. XPM)

by extending Eq. (2.6) to a generalized NLSE (GNLSE) [3]

$$\frac{\partial A}{\partial z} = -\frac{\alpha}{2}A + \sum_{k\geq 2}\frac{i^{k+1}}{k!}\beta_k\frac{\partial^k A}{\partial t^k} + i\gamma\left(1 + i\tau_{shock}\frac{\partial}{\partial t}\right) \cdot \left(A\int_{-\infty}^{+\infty} R(t') \times |A(z, t-t')|^2 dt'\right), \tag{3.1}$$

where α is the absorption coefficient, the term with β_k describes the higher order dispersion, and γ is the SPM-coefficient. The second term describes the effect of SPM and the effect of self-steepening with the parameter $\tau_{shock} = 1/\omega_0$. Furthermore, Raman scattering is considered by the electronic response function $R(t) = (1 - f_R)\delta(t) + f_R h(t)$. A typical value for the parameter f_R is 0.18 [3]. Eq. (3.1) is usually solved numerically by the split-step Fourier method [28]. In the following, a qualitative description of the dynamics of supercontinuum generation in the femtosecond regime based on ref. [3] is given. The temporal dynamics of the optical spectrum and the pulse envelope are visualized for the propagation of a pulse with a duration of 50 fs along 15 cm of PCF (Fig. 3.3). In the femtosecond regime, supercontinuum generation is dominated by soliton dynamics. This is in contrast to supercontinuum generation with longer pulses where Raman scattering, four-wave-mixing, and modulation instabilities play the most important role. At the initial stage of supercontinuum generation, that is approximately during propagation over the first 0.5 cm, a strong symmetrical broadening of the spectrum as well as a temporal compression happens due to SPM. Finally a usually higher order soliton is formed. After approximately 1 cm of propagation along the waist region, the spectral broadening becomes asymmetric with distinct peaks on the short- and the long-wavelength side. These peaks are caused by the fission of the higher order soliton. This can be caused by Raman scattering or by perturbations due to higher order dispersion. Which effect dominates is dependent on the input pulse width. For pulses with $\tau > 200$ fs, Raman scattering is the main reason; for pulses with $\tau < 50$ fs, nearly only higher-order dispersion causes the fission. For pulse-widths in between, the contributions are approximately equal. The higher-order soliton breaks up into several fundamental solitons, or solitons of lower orders. Hereby the solitons experience a red-shift, the so-called soliton self-frequency shift [5, 4]. Simultaneously, narrow-band blue-shifted light is emitted at the normal GVD regime (dispersive wave). The exact wavelength of this dispersive wave is determined by a phase-matching condition [3]

$$\frac{\omega_{dw}}{v_{g,s}} - \frac{\omega_s}{v_{g,s}} = \beta(\omega_{dw}) - \beta(\omega_s) - (1-f_R)\cdot\gamma\cdot P_s, \tag{3.2}$$

where ω_{dw} and ω_s are the frequencies of the dispersive wave and the soliton, respectively. $\beta(\omega_{dw})$ and $\beta(\omega_s)$ are the propagation constants and $v_{g,s}$ is the group velocity at the soliton wavelength. Fig. 3.2 b) shows a calculation of the phase difference between a fundamental soliton and the dispersive wave for a tapered fiber with a diameter of 2.2 μm and an input power of 100 mW [31]. One recognizes that for longer soliton wavelengths (here at 950 nm) the phase-matching condition leads to shorter dispersive wavelengths. This means that the greater the difference between the pump wavelength to the zero-dispersion wavelength (see Fig. 3.2 (a)), the further blue-shifted is the non-solitonic peak. The break-up into individual pulses can also be seen on the pulse envelope (Fig. 3.3 a)). While propagating along the waist region, the fundamental solitons undergo a further red-shift due to Raman-scattering which broadens the spectrum at the long-wavelength side [5]. However, after the fission of higher-order solitons, nearly no further broadening on the blue-side of the spectrum appears. Furthermore, over the whole waist region, different frequency components can mix by four-wave-mixing processes. For example, the dispersive wave and fundamental solitons can mix with each other by cross-phase modulation (XPM).

In Fig. 3.4 different measured spectra of supercontinua are shown. Fig. 3.4 a) shows spectra that were obtained with two different fibers with a waist diameter of 3.8 μm (black line) and 5 μm (gray line), respectively. The red line indicates the position of the pump wavelength. The output power was held constant at approximately 1.2 W in each case. In this case, one can clearly see that the fiber with the thinner diameter generates a peak of the dispersive wave at shorter wavelengths than the thicker fiber. The position of the dispersive wave peak is indicated by the marker (A). While the position of the dispersive wave depends on the waist diameter, the red-wavelength side of both spectra in Fig. 3.4 a) are equally broad (B). Also the peaks of the red-shifted solitons are visible as marked by (B) and (C). Furthermore, the gap between the peak of the dispersive wave and the broad infrared part can be noticed (D). The infrared part spans approximately from the pump wavelength to the red-wavelength side of the spectrum. In order to additionally demonstrate the influence of the pump power, Fig. 3.4 b) shows three spectra that were measured with the same fiber with a diameter of 2.7 μm, but for different pump powers, resulting in different output powers. Here, one can see that the red-wavelength side broadens to a greater degree with increasing input powers but the position of the dispersive wave peak remains approximately constant. Even if the position of the dispersive wave peak does not change, the peak becomes more pronounced for higher powers. For the highest output power of 290 mW (black curve), one can even recognize a second blue-shifted peak emerging next to the dispersive wave peak. This might be due to the fact that with higher pump powers the order of the initial soliton increased. During fission of this increased order, a soliton was generated that further red-shifted due to the soliton-self-frequency shift than in the cases of a lower pump power. Therefore, according to the phase-matching condition Eq. (3.2), an additional further blue-shifted peak could be generated.

The mechanisms of supercontinuum generation are different for input pulses in the femtosecond regime and for longer pulse durations. Since incoherent Raman scattering plays a less important role in the femtosecond regime, the phase relationship between spectral components as well as the coherence of successive pulses is preserved [3, 16]. Furthermore, the decoherence due to input noise turned out to be less strong the shorter the duration of the input pulses [3]. This is important, for example, if spectrally filtered sections of supercontinuua need to be temporally compressed before being applied in an experiment.

Chapter 4
Solitary mode-locked laser oscillators

A laser oscillator can be operated in three principal operating regimes. In the continuous wave (cw) regime, the field envelope $A(t)$ is constant, whereas in the other two regimes, Q-switching and mode-locking, the field envelope is modulated and the laser output power is emitted in short pulses. The time span between two successive pulses can be huge compared to the pulse duration. In addition to operation in one of these ideal regimes, a laser oscillator can also operate in a combination of different regimes. For instance, a mode-locked pulse train can be combined with a cw background, or a laser can operate in the Q-switched mode-locking regime where the pulse train with short mode-locked pulses is additionally modulated with giant pulses due to Q-switching. In this Q-switched mode-locked regime, a laser oscillator does not emit a pulse train of identical pulses, but successive pulses vary in amplitude. Ideal mode-locking with a pulse train of identical laser pulses and no background is also referred to as cw mode-locking.

The physical reasons which force a laser to operate in the Q-switching or in the mode-locked regime are completely different. While Q-switching is induced by an amplification of relaxation oscillation peaks due to a modulation of the resonator quality, hence the name in the mode-locked regime the laser pulses are formed by the superposition of many longitudinal modes with fixed phase relationship between them.

In this work, we decided to focus on laser oscillators operating in a regime called solitary mode-locking [34, 35]. Part of the mode-locking mechanism hereby is a semiconductor saturable absorber mirror (SESAM) [36, 37, 38, 39, 40]. The regime of solitary mode-locking is particularly suitable for the purpose of femtosecond supercontinuum generation because such laser oscillators can be operated in a very stable condition, which is in contrast to Kerr-lens mode-locked lasers for example. Furthermore, such laser oscillators can be self-starting, i.e. no external disturbance is necessary to initiate mode-locking. The mode-locking mechanism in this regime is an interaction between a slow saturable absorber, the SESAM, and pulse shaping by soliton formation. Since the absorber relaxation times of SESAMs, which are typically on the order of a few ten ps, are long compared to pulse durations of less than 200 fs, a SESAM can only initiate and stabilize mode-locking. The pulse formation in the solitary mode-locked regime is achieved by the formation of optical solitons due to insertion of anomalous GDD into the laser resonator. However, since the mode-locking driving force is provided by the SESAM, the laser resonator can be designed to operate in the middle of a stability regime such that the resonator properties are similar for cw and pulsed operation. One challenge in the design of laser resonators for solitary mode-locked lasers is the use of a slow saturable absorber. Besides stabilizing the mode-locked operation, the introduction of such a slow saturable absorber can also introduce a tendency to Q-switching. Therefore, in order to find an operating regime where the mode-locking driving force is sufficiently strong but simultaneously Q-switching is suppressed, it is important to determine a suitable set of laser parameters. Therefore necessary tools can be derived by regarding the dynamics of laser oscillators.

4.1. Dynamics of mode-locked laser oscillators

4.1.1. Laser rate equations and master equation of mode locking

In the following, a description of the dynamics of mode-locked laser oscillators is presented. Based on this model, the necessary design parameters for mode-locked lasers can be derived. The model is mostly based on the theory developed by H.A. Haus [41], and extended with regard to passive solitary mode-locking with saturable absorbers by Kärtner et al. [37] and Hönninger et al. [42].

A very practical model to describe the dynamics of a mode-locked laser can be found by considering the (resonator internal) average laser power $P(T)$, the gain per one resonator round-trip $g(T)$, the time-dependent losses $q(T, t)$, and the exact shape of the pulse envelope $A(T, t)$. Here, as in the description of pulse propagation with the NLSE (Eq. 2.6), the parameter T is the global time, whereas t is the retarded time with respect to the peak of the laser pulse in the resonator. Time-

4.1. Dynamics of mode-locked laser oscillators

dependent power losses $q(T,t)$ are necessary in order to provide a state in which mode-locked operation is energetically favorable compared to cw operation. This can experimentally be achieved by various active or passive methods. Examples for a passive methods are saturable absorbers which bleach at high peak intensities. Therefore intensities during pulsed operation can bleach such an absorber while it absorbs at cw operation. Another method would be to design a laser resonator in such a way, that the laser mode and pump mode overlap is better when the Kerr-lens is sufficiently strong. The three variables $P(T)$, $g(T)$, and $q(T,t)$ can be described by laser rate equations. The time-dependent envelope $A(T,t)$ can be modeled by a combination of the NLSE for pulse propagation (Eq. 2.6) and the laser rate equations. The latter step leads to the master equation of mode-locking as developed by H.A. Haus [41]. The average power P and the pulse envelope $A(t)$ are related to each other by the energy per one pulse E_P, as well as the time for one resonator round-trip T_R. The pulse energy can either be expressed by $E_P = P \cdot T_R$ or by $E_P = \int |A(t)|^2 dt$. Here, $A(t)$ needs to be normalized with respect to the time-dependent power, i.e. $|A(t)|^2 = P(t)$. The rate equations for the particular case with $q(t)$ realized by a saturable absorber can be written as follows [42]:

$$\frac{dP}{dT} = \frac{g-l-q}{T_R} \cdot P, \qquad (4.1)$$

$$\frac{dg}{dT} = \frac{g-g_0}{\tau_L} - \frac{P}{E_{sat,L}} \cdot g, \qquad (4.2)$$

$$\frac{dq}{dT} = \frac{q-q_0}{\tau_A} - \frac{P}{E_{sat,A}} \cdot q, \qquad (4.3)$$

where l are the unsaturable intensity losses per round-trip, τ_L and τ_A are the upper state lifetimes of the gain medium and the saturable absorber, respectively, q_0 is the saturable but non-saturated loss of the absorber, g_0 is the small-signal gain which is proportional to the emission cross section as well as to the pump power, $E_{sat,L} = h\nu/(m\sigma_L)A_{eff,L}$ is the saturation energy of the gain medium, and $E_{sat,A} = F_{sat,A}A_{eff,L}$ is the saturation energy of the saturable absorber. $A_{eff,L}$ and $A_{eff,A}$ are the effective mode areas in the gain medium and on the saturable absorber, respectively, m determines the number of passes through the gain medium, σ_L is the corresponding emission cross-section, and ν is the laser frequency. The effective mode areas A_{eff} are given by $A_{eff} = \pi w^2$ with the $1/e^2$ Gaussian beam radius w. The model given in Eq. (4.3) is valid for saturable absorbers, for example doped crystals or SESAMs. The pulse envelope $A(T,t)$ can described by an extension of the NLSE (Eq. 2.6), the so-called master-equation of mode-locking [41, 37]:

$$T_R \frac{\partial}{\partial T} A(T,t) = \left(g(T) - l + D_{g,f} \frac{\partial^2}{\partial t^2} + i \frac{k_2}{2} \frac{\partial^2}{\partial t^2} - q(T,t) - i\delta |A(T,t)|^2 \right) A(T,t). \quad (4.4)$$

The term with the constant $D_{g,f} = 1/\Omega_g^2 + 1/\Omega_f^2$ takes the gain and filter losses due to the broad optical spectrum into account. Ω_g and Ω_f are the FWHM gain and filter bandwidth of the gain medium and other wavelength filtering elements, respectively. $\delta = \gamma \cdot L = \frac{\omega_0 \cdot n_2}{c \cdot A_{eff,L}} L$ is a modified SPM coefficient. L is the total propagation length inside the gain medium (or the medium in which SPM appears) per one round-trip and T_R is the time of one resonator round-trip.

4.1.2. Suppression of Q-switching

When a slow absorber is inserted into a laser oscillator, the formation of giant pulses can appear because of Q-switching. This happens if the absorber is not sufficiently saturated and thus peaks in the relaxation oscillations of the laser power can be amplified. Therefore, in order to achieve pure cw mode-locking, one needs to find a regime in which instabilities due to Q-switching are suppressed. Such a criterion was derived by Hönninger et al. [42] by a stability analysis of the linearized rate Eqs. (4.3) for the particular case of using SESAMs as saturable absorbers. According to this criterion, Q-switching is suppressed if the saturable reflectivity change ΔR (modulation depth) of the employed SESAM is kept below a certain limit:

$$\Delta R < \frac{E_P}{F_{sat,A} A_{eff,A}} \left(\frac{T_R}{\tau_L} + \frac{E_P}{E_{sat,L}} \right). \quad (4.5)$$

4.1.3. Initiation and self-starting of mode-locking with saturable absorbers

If pure cw mode-locking should be initiated by a SESAM and Q-switching needs to be suppressed, the modulation depth ΔR must be kept lower than the limit of Eq. (4.5). Even in cases where the modulation depth is chosen sufficiently low to suppress Q-switching, it can still be high enough to excite several longitudinal modes and lead to mode-locking. A criteria for a saturable but non-saturated absorption q_0 that is high enough to initiate mode-locking was derived by Kärtner et al. by calculating the build-up time T_{mod} for two neighboring modes [37]. Mode-locking is possible if the build up time T_{mod} is positive. Hence, if the following relation is fulfilled:

$$\frac{1}{T_{mod}} = \left(\frac{q_0}{(1+P/P_A)^2 + (2\pi m T_A)^2} \frac{P}{P_A} - \frac{2g_0}{(1+P/P_L)^2 + (2\pi m T_L)^2} \frac{P}{P_L} \right) \bigg|_{cw} > 0, \tag{4.6}$$

with the average laser power P, the saturable but non-saturated absorption q_0, the small-signal gain g_0, the cavity mode m times $2\pi/T_R$ away from the center mode, the upper state life-times T_L and T_A of the gain medium and the saturable absorber, normalized to the round-trip time T_R, and the saturation power P_L and P_A of the gain medium and the absorber, respectively.

In the experiment, this condition is hard to determine because the gain profile and number of modes needs to be known exactly. Therefore, it is more realizable to empirically choose the lowest possible saturable absorption for which mode-locking with the intended pulse width can still be initiated. Using the lowest possible modulation depth is beneficial because the non-saturable losses increase with higher saturable absorption, for example due to increased two-photon absorption in the thicker quantum wells [43].

4.2. Solitary mode-locking with semiconductor saturable absorber mirrors (SESAMs)

Gain media that are very efficient often possess long upper state lifetimes. Therefore only low modulation depths can be used to achieve pure cw mode-locking. This is the reason why such solid state lasers could be efficiently cw mode-locked for the first time by employing semiconductor based saturable absorbers (SESAMs) [40]. Before the realization of SESAMs, commonly used slow saturable absorbers showed too high modulation depths and cw mode-locking could only be achieved by fast absorbers, for instance with Kerr-lens mode-locking. Today, usually anti-resonant low-finesse SESAMs are used for femtosecond lasers with medium average powers of a few Watts [38]. Such a low-finesse A-FPSA SESAM consists of a sandwich of a highly reflective bottom Bragg reflector, one or several quantum-well absorber layers, spacer layers, and a partially reflective top Bragg reflector.

Fig. 4.1 shows a SESAM structure for an operating wavelength of ~ 1 μm. The bottom reflector and the absorber layers are realized as semiconductor Bragg reflectors. For instance for a laser wavelength around 1 μm the Bragg reflector consists of GaAs and AlAs quarter wave layers. The absorber layer is realized by thin InGaAs layers and spacer layers are employed to achieve anti-resonance in the quantum

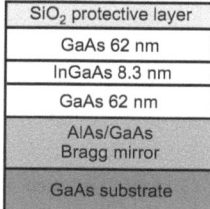

Figure 4.1.: Structure of a low-finesse anti-resonant SESAM for an operating wavelength around $\lambda_L \approx 1$ μm according to Refs. [38, 44].

wells. The top reflector can either be a partially reflective Bragg reflector of dielectrics such as SiO_2, TiO_2, or AlO_2 quarter wave layers, or employ the Fresnel reflection between air and the semiconductor or a SiO_2 protection layer. SESAMs are usually designed in such a way as to achieve an anti-resonant field in the absorber layer which leads to a broadband operating regime. Furthermore, low modulation depths can be achieved by this way. The saturation fluence $F_{sat,A}$ is determined by the reflectivity of the top layer, as well as the properties of the absorber layers. For increasing laser power, the heat load deposited in low-finesse SESAMs due to non-saturable losses also increases. In this case, high-finesse SESAMs with nearly highly reflective top mirrors might provide a suitable alternative.

One difficulty when using SESAMs to achieve femtosecond laser pulses is that the recovery time of the absorber is much longer than the width of the generated laser pulses. Even though the recovery times can be reduced to only 3-10 ps by sophisticated fabrication techniques [37, 45, 46, 47], such SESAMs cannot form laser pulses shorter than the recovery time. In this case, one needs to employ the possibility of solitary mode-locking [34, 35, 45]. If femtosecond laser pulses should be generated in this regime, the SESAM only initiates and stabilizes mode-locking, however, the laser pulses are formed by the interaction between SPM and GVD. Only such pulses which fulfill the condition for a fundamental soliton can propagate many times in the laser resonator without losses. All other pulses experience a spectral broadening which reduces the gain. Furthermore, these pulses temporally broaden until they form a vanishing background. Usually the intra-cavity GDD per round-trip must be adjusted to support a given pulse width at a certain laser power [35]:

$$|D_2| = \frac{\tau_{FWHM} \cdot \delta \cdot E_P}{4 \cdot 1.76}, \qquad (4.7)$$

where D_2 is the intra-cavity GDD per one round-trip, τ_{FWHM} is the FWHM pulse widht of the solitons, $\tilde{\delta}$ is the modified SPM coefficient $\delta = \gamma \cdot L = \frac{\omega_0 \cdot n_2}{c \cdot A_{eff,L}} \cdot L$, and E_P is the pulse energy which is related to the average power P and round-trip time T_R by $E_P = P \cdot T_R$.

4.3. Pulse break-up and the B-integral

Especially for high laser powers and long propagation lengths in the gain medium, as in slab lasers, the nonlinear effects due to SPM and self-focusing can become too strong and no fundamental soliton can evolve anymore. A limit for maximum nonlinear effects can be given in terms of the so-called B-integral [48]

$$B = \frac{2\pi}{\lambda} \int_0^L n_2(z) \cdot I(z) dz, \qquad (4.8)$$

where n_2 is the nonlinear index, and L is the total resonator length and the term $I(z)$ denotes the laser intensity $I(z) = \frac{|A(z)|^2}{A_{eff,L}}$. Usually, n_2 is only non-zero in the gain medium. A general criterion for high-power lasers states that the B-integral must be kept below $B \leq$3-5 to avoid distortions due to SPM or self-focusing [48].

Chapter 5
Resonator design for diode-pumped femtosecond slab lasers

5.1. Numerical model of the thermal effects in longitudinally diode-pumped gain media

Longitudinal diode-pumping of laser gain media usually results in a highly non-uniform temperature distribution. On one hand, the heat that is generated by the absorption of pump light is due to the quantum defect. On the other hand, it can be generated by non-radiative decay processes or other effects, for instance by re-absorption of fluorescence or laser light. The resulting inhomogeneous temperature distribution is the origin of several successive effects which affect the laser beam that propagates through the gain medium.

In order to support the laser design and the interpretation of experimental results, it is important to have a theoretical model and description of the thermal effects that appear in the gain medium. Such modeling of thermal effects has been a topic of research since the development of Nd:YAG lasers in the 1960's. For example Osterink et al. for the first time could increase the average power of Nd:YAG lasers to over 1 W with operation in the TEM$_{00}$ mode by considering the thermal lens [49]. For high-power lasers, Koechner could develop an analytical solution for the temperature distribution and the focal length of the thermal lens in flash-lamp pumped Nd:YAG rods [50]. With the rise of solid state lasers end-pumped by laser diodes, also the thermal effects in longitudinally diode-pumped lasers were investigated [51, 52, 53, 54]. However, analytical solutions can only be found for simple configurations, i.e. highly symmetric geometries and beam-profiles such as cylindrical rods or slabs of isotropic materials and Gaussian or top-hat beam-profiles.

Chapter 5: Resonator design for diode-pumped femtosecond slab lasers

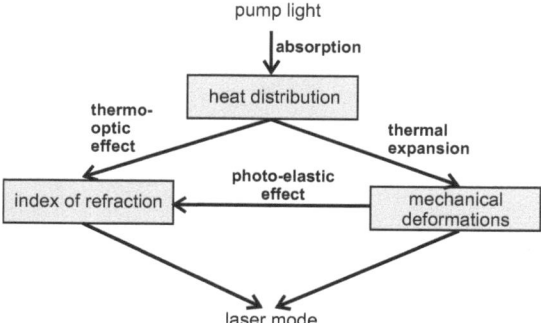

Figure 5.1.: Schematic diagram of the effects caused by heating of the gain media due to pump light absorption.

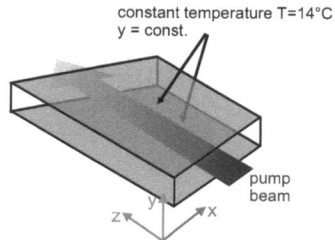

Figure 5.2.: The flat Brewster-cut Yb:KGW crystal is pumped collinear with the laser beam. The two red surfaces are kept at a constant temperature of $T = 14°C$ and are mechanically fixed which is modeled by keeping the y-positon of these two planes constant.

For anisotropic gain materials, asymmetric geometries, or complex beam profiles, solutions can only be obtained numerically.

To numerically calculate the thermal effects in this work, the involved processes were studied by a finite element analysis (FEA). The simulations did consider two physical models, heat conduction and thermal expansion plus the thereby induced mechanical stresses. All effects, the dependencies amongst them, and their influence on the laser beam are visualized in Fig. 5.1. The basis of all related effects is the development of an inhomogeneous temperature distribution. On one hand, this affects the laser beam directly because the index of refraction is temperature dependent (thermo-optic effect). On the other hand, the temperature gradients in combination with thermal expansion of the crystal lead to a bulging of the crys-

5.1. Numerical model of the thermal effects in longitudinally diode-pumped gain media

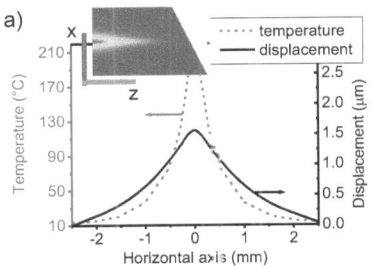 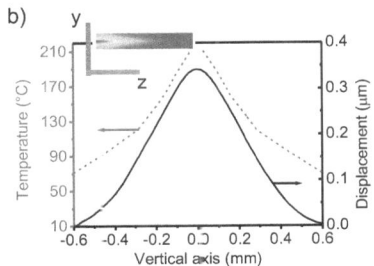

Figure 5.3.: Calculated temperature distribution (red dashed line) and displacement (black solid line) along a horizontal (a) and a vertical (b) cross section over the center of the flat crystal side. The simulations were done for the longitudinally pumped Yb:KGW crystal with 12 W pump power, a Gaussian beam profile with a diameter of $\sim 400 \times 200~\mu m^2$, and $M^2 = 50 \times 1.2$. The insets show the temperature distribution in two perpendicular planes through the crystal center.

tal surface which can act as a lens. Furthermore, mechanical stresses appear in the medium which can also influence the laser mode due to an additional stress-dependence of the refractive index (photo-elastic effect).

The setup and gain medium geometry which was used is schematically shown in Fig. 5.2. The gain medium was modeled with a three-dimensional flat-Brewster cut geometry. The material parameters of the anisotropic Yb:KGW were entered along each axis according to Chapter 7.3.5. The pump beam enters the crystal on the flat side and decays exponentially over the propagation length. The assumption of an exponential pump light decay is only valid if no saturation effects occur, what we estimated to be a reasonable assumption for our case. As boundary conditions, the red surfaces of the crystal were held at a constant temperature of $T = 14°C$ and physically contacted to the copper heat sink through indium foil. To model this contact, the surfaces were assumed to be mechanically fixed in the y-direction, but allowed to slide along the x and z-direction to describe the embedding of the crystal into soft indium foil. The stationary heat distribution was calculated by the heat equation

$$-\nabla\big[k(T)\nabla T(\mathbf{r})\big] = Q(\mathbf{r}), \tag{5.1}$$

where $T(\mathbf{r})$ is the spatial temperature distribution, $Q(\mathbf{r})$ is the heat source, and $k(T)$ is the temperature dependent thermal conductivity. In the mechanical part of the simulation, the overall stresses and the deformation of the gain medium were calculated. While the deformation (displacement) is calculated with the Hook's law $\sigma = E\epsilon$, the overall stresses σ in the gain medium are caused by volume forces $\mathbf{F}$ ($[\mathbf{F}]=\mathrm{N/m}^3$) and can be expressed by the equation

$$-\nabla\sigma = \mathbf{F}. \tag{5.2}$$

As input to Eq. 5.2, the volume forces caused by the stress due to thermal expansion σ_{th} were used, and the overall stresses were obtained by solving Eq. (5.2). To combine both models, heat conductivity and the mechanical part, the volume forces need to be related to the temperature distribution. This can be achieved by incorporating the thermal expansion Δl into the generalized Hook's law

$$\sigma_{th} = E \cdot \epsilon_{th} = E \cdot \alpha \cdot \Delta T, \tag{5.3}$$

with the Young's modulus E, the strain $\epsilon_{th} = \Delta l/l$, the thermal expansion coefficient α, and the temperature difference ΔT. Therefore, the volume forces caused by thermal expansion can be expressed by

$$\mathbf{F} = -\mathbf{E} \cdot \alpha \cdot \nabla \mathbf{T}, \tag{5.4}$$

whereas α is the thermal expansion coefficient tensor. With the FEA software (Comsol Multiphysics), initially the pump light absorption is calculated. During this step, the specific beam profile as well as the laser efficiency can be taken into account. Based on this first result, the temperature distribution inside the gain medium is determined. With the temperature distribution, the stresses and the displacement can then be calculated. All steps need to be repeated iteratively until a self-consistent solution is found.

Fig. 5.3 shows the result for the temperature distribution (red dashed line) and displacement (solid black line) along a horizontal and a vertical cross-section over the center of the flat side of the crystal. The simulations were done for the longitudinally pumped Yb:KGW crystal with 12 W pump power of a broad area emitter, a Gaussian beam profile with a diameter of $\sim 400 \times 200$ $\mu\mathrm{m}^2$, and a beam quality of $M^2 = 50 \times 1.2$ as described in Chapter 7.3.5. The insets show the temperature distribution in two perpendicular planes through the crystal center. The temperature distribution shows approximately a parabolic shape in the proximity of the laser beam ($x = $ -0.2...+0.2 mm, $y = $ -0.1...+0.1 mm), and a logarithmic decay further away from the center. Such a behavior would be expected from an analytical solution of the temperature distribution in a cylindrical rod end pumped with a laser

5.1. Numerical model of the thermal effects in longitudinally diode-pumped gain media

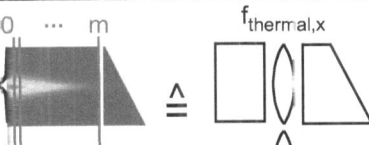

Figure 5.4.: To retrieve the focal length of the thermal lens, the bulged surface is considered as parabolic lens and the gain medium is divided into m thin slices which are each treated as a gradient index lens. The overall ray-matrix of the left system is compared to an equivalent system with two cylindrical lenses inside the laser crystal. The equivalent lenses are calculated in such a way, that the overall ray matrices of both optical systems are equal.

diode with a top-hat beam-profile [55]. A similar behavior can be observed for the displacement. Therefore, in proximity to the pump beam, both effects can be fitted with parabolic functions and the thermal effects can be regarded as a parabolic lens [54]. Deviations from the parabolic behavior lead to higher order aberrations.

5.1.1. Retrieval of the focal length of the thermal lens

To retrieve the focal length of the thermal lens, the data of such a FEA-simulation was analyzed. Each calculation needs to be done for the tangential (xz) and the sagittal-plane (yz), respectively. First, the bulging of the surface is fitted with a parabolic function with curvature $A_{x/y}$. With the index of refraction of the gain medium n_0, the focal length of this contribution can be calculated to

$$f_{bulge,x/y} = \frac{1}{(n_0 - 1) \cdot 2A_{x/y}}. \qquad (5.5)$$

This effect can also be described by the ray-matrix formalism [48]. The parabolic lens due to bulging of the crystal surface, is described by the matrix $M_{bulge,x/y}$ which is the matrix of a thin lens with the focal length $f_{bulge,x/y}$. If one divides the gain medium into m thin slices of thickness d, the temperature inside these slices is nearly constant along the propagation direction, but shows an approximately parabolic lateral temperature profile. With the thermo-optic parameter $\partial n/\partial T$, this temperature profile can be transformed into a distribution of the index of refraction $n(x, y) = n_0 + \partial n/\partial T \cdot (T(x, y) - T_0)$. Each of these thin slices can then be modeled as a gradient index lens with a ray-transfer matrix $M_{i,x/y}$ [56]

$$M_{i,x/y} = \begin{pmatrix} \cos b_{i,x/y} & \left(\frac{d}{n_0}\right)\left(\frac{\sin b_{i,x/y}}{b_{i,x/y}}\right) \\ -\left(\frac{n_0}{d}\right) b_{i,x/y} \sin b_{i,x/y} & \cos b_{i,x/y} \end{pmatrix}, \quad (5.6)$$

whereas n_0 is the index of refraction, and the coefficients $b_{i,x/y}$ are determined by $b_{i,x/y} = d\sqrt{\frac{2B_{i,x/y}}{n_0}}$. $B_{i,x/y}$ are the curvatures of the parabolic fit-functions to the refractive index profiles. With the ray-matrix of an infinitely thin slice of a flat Brewster-cut crystal $M_{f-B,x/y}$, the overall ray-matrix for propagation through the gain medium can be calculated by

$$M_{tl,ges,x/y} = M_{f-B,x/y} \cdot \prod_{i=1}^{m} M_{i,x/y} \cdot M_{bulge,x/y}. \quad (5.7)$$

Usually the laser resonator is modeled by the ray-transfer formalism with discrete elements. Therefore it is beneficial to reduce the huge expression of Eq. 5.7 to two equivalent thin cylindrical lenses at the center of the laser crystal (see Fig. 5.4). The overall ray-matrix of such an equivalent system would be the product of the matrix of an rectangular block $M_{1,x/y}$, the matrix of the thin cylindrical lens $M_{thermal,x/y}$, and the matrix for a flat Brewster-cut block $M_{2,x/y}$. By comparing this equivalent matrix with Eq. 5.7

$$M_{tl,ges,x/y} = M_{2,x/y} \cdot M_{thermal,x/y} \cdot M_{1,x/y},$$

the ray-matrix for the thermal lens can be obtained

$$M_{thermal,x/y} = M_{2,x/y}^{-1} \cdot M_{tl,ges,x/y} \cdot M_{1,x/y}^{-1}. \quad (5.8)$$

Except for small numerical artifacts, the matrix $M_{thermal,x/y}$ corresponds to the matrix of a thin lens. Hence, the focal length $f_{thermal,x/y}$ of the thermal lens can be retrieved from this matrix.

5.2. Simulation of the resonator internal beam waist

When designing a laser resonator, it is important to know if the resonator is stable in the first place, as well as the diameter of the laser mode at certain positions. For instance, the mode diameter inside the laser medium or on the SESAM. The calculation of these diameters can also be done with the ray-transfer matrix method. However, for this purpose, the method needs to be applied to Gaussian beams and the complex q-parameter $\frac{1}{q(z)} = \frac{1}{R(z)} - i\frac{\lambda}{\pi w(z)^2}$, where $R(z)$ is the phase front radius

5.2. Simulation of the resonator internal beam waist

of curvature and $w(z)$ is the $1/e^2$ beam radius [48]. When propagating through an optical element with a ray-matrix $M = \begin{pmatrix} A & B \\ C & D \end{pmatrix}$, the q-parameter before the element q_{in} transforms into the parameter q_{out} after the element following the relation

$$q_{out} = \frac{Aq_{in} + B}{Cq_{in} + D}. \tag{5.9}$$

A laser resonator can be modeled by the ray-matrix product of all optical elements along one round-trip. Hence, the overall ray-matrix M_{total} of a laser resonator with n optical elements can be described by

$$M_{total} = M_1 \cdot M_2 \cdots M_{n-1} \cdot M_n \cdot M_{n-1} \cdots M_2. \tag{5.10}$$

For a stable laser resonator, the absolute value of the eigenvalue λ of M_{total} needs be to be smaller than one, and the determinant of M_{total} needs to be between ± 1. Regarding these conditions, the general stability limit of a laser resonator is given by

$$\left| \frac{A_{total} + D_{total}}{2} \right| \leq 1. \tag{5.11}$$

The initial q-parameter q_0 for the laser eigenmode at the surface of one of the two end-mirrors can be derived by noting that the q-factor should be preserved after one round-trip, i.e. $q_0 = \frac{A_{total} q_0 + B_{total}}{C_{total} q_0 + D_{total}}$. The initial q-parameter q_0 is given by

$$\frac{1}{q} = \frac{D_{total} - A_{total}}{2B_{total}} \pm \frac{1}{B_{total}} \sqrt{\frac{1}{4}(A_{total} + D_{total})^2 - 1}, \tag{5.12}$$

where only the negative root can be considered in order to achieve a real value for $w(z)$ on the end-mirror. The beam radius $w(z)$ along the whole resonator length can then be calculated by successively applying the matrices M_i and propagation matrices on q_0. The thermal effects can be considered by inserting additional optical elements into the simulation, for instance two thin cylindrical lenses with the focal length $f_{thermal,x/y}$ for the thermal lens. A list of all used ray-matrices is given in appendix C.

In mode-locked operation, the peak intensities inside the gain medium are sufficiently high to excite an additional Kerr-lens which influences the laser resonator. This Kerr-lens is caused by an intensity dependent refractive index described by the nonlinear index n_2. This effect is also known as self-focusing or the optical Kerr effect. Since the laser mode possesses a lateral beam profile, for instance a Gaussian intensity distribution, also a lateral refractive index distribution is created. In the region near the center of the beam profile, this distribution can be fit with a

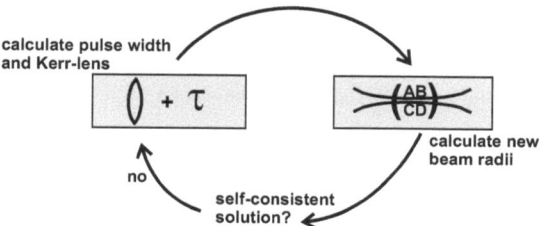

Figure 5.5.: Iterative algorithm to calculate the beam radius inside the laser resonator, as well as the expected pulse width.

parabolic function. Hence, the effect on the laser beam can be modeled by a gradient index lens with the ray matrix M_{KL} [57, 48]

$$M_{KL} = \begin{pmatrix} \cos\gamma d & \frac{1}{n'_0\gamma}\sin\gamma d \\ -n'_0\gamma\sin\gamma d & \cos\gamma d \end{pmatrix}, \quad (5.13)$$

where d is the propagation length in the Kerr-medium, n'_0 is the effective refractive index

$$n'_0 = n_0 + n_2\frac{2\hat{P}}{\pi w^2}, \quad (5.14)$$

and the parameter γ is describe by

$$\gamma = \frac{1}{w^2}\sqrt{\frac{8n_2\hat{P}}{n'_0\pi}}. \quad (5.15)$$

The variable w is the $1/e^2$ beam radius inside the Kerr-medium, and $\hat{P}$ is the peak power. The peak power is related to the pulse energy E_P and the FWHM pulse width τ_{FWHM} by

$$\hat{P} = \frac{1.76 \cdot E_P}{2 \cdot \tau_{FWHM}} = 0.88 \cdot \frac{E_P}{\tau_{FWHM}}. \quad (5.16)$$

When modeling the propagation in the laser resonator for mode-locked operation, the propagation through the gain medium must be described by a ray matrix for propagation in addition with self-focusing M_{KL}. In that case, an iterative algorithm needs to be employed (Fig. 5.5). In the first step, the beam waist inside the 'cold' laser resonator is calculated. Based on these results, the pulse width can be estimated by Eq. 4.7. With an adapted ray-matrix of the Kerr-lens, the beam waist inside the resonator can be re-calculated. This procedure needs to be repeated until a self-consistent solution is found, or the iteration can be stopped, if the resonator becomes unstable. In this work, this algorithm was realized by a MATLAB code.

5.2. Simulation of the resonator internal beam waist

The laser crystal was modeled by two lumped elements of half the crystal length. This proved to be too coarse for the calculation of some resonators, especially for high laser intensities and strong nonlinearities. Thus, the code could be further improved by modeling the laser crystal by smaller steps.

Chapter 6
Compact 20 MHz femtosecond supercontinuum system based on an Yb:glass oscillator

An approach for a very compact and portable femtosecond supercontinuum source is described. This novel source possesses similar properties as systems based on Ti:sapphire lasers [2, 7], and can therefore provide a flexible and very rugged alternative for these setups. The first part of the chapter describes a novel design for a directly diode-pumped femtosecond Yb:glass laser oscillator. By employing a Herriott-type multi-pass cell, the laser footprint was fit to a size comparable to that of fiber lasers. The second part of this chapter shows experimental results of supercontinuum generation with this laser in tapered fibers as well as in PCF. Furthermore, experiments to study important parameters with regard to practical applications, such as the intensity noise of supercontinua, the influence of fiber pig-tails, and the influence of a pulse chirp on the spectra, are presented. The results of these experiments show that femtosecond supercontinuum sources based on Ytterbium lasers provide a more practical and rugged solution compared with sources based on Ti:sapphire lasers. The properties which are investigated can also serve as an example for results obtainable with other Ytterbium lasers with similar properties.

6.1. Current state of research

In recent years it has been shown that supercontinua that are generated by pulses shorter than a limit of approximately 200 fs are mostly based on coherent processes, and, therefore, possess some interesting properties as discussed in section 3 [3, 16]. At first, experiments with femtosecond supercontinua were realized mostly with Ti:sapphire lasers [2, 7]. Nearly all properties of these supercontinua were investigated based on pump wavelengths around 800 nm. However, to implement such lasers into other experiments and in harsher environments, it would be beneficial to provide more compact, stable, and hands-off systems. In order to pursue this goal, Teipel et al. investigated possibilities of femtosecond supercontinuum generation with Yb:glass lasers with wavelengths around 1 μm [13]. Ytterbium lasers are particularly well suited pump sources for this purpose since they can be directly diode-pumped, yet are capable of simultaneously generating femtosecond laser pulses with the desired pulse widths below 200 fs [58, 59]. These lasers also offer the potential to build very compact systems. A possible alternative to solid-state lasers can be provided by fiber lasers. With Yb:silica fiber lasers it is possible to achieve even sub-100 fs pulses [60, 61]. However, nonlinearities and dispersion play an important role for femtosecond fiber lasers and need to be taken into account which makes the experimental setups more complex. For instance, additional external dispersion control or pulse compression is necessary.

The goal of this work was to investigate an approach to further reduce the size of such compact and rugged setups for femtosecond supercontinuum generation. Based on the former work of Teipel et al. [13], we decided on a setup based on an Yb:glass laser oscillator and tapered fibers. Additionally the possibilities of PCF were investigated as well. Furthermore, the experiments aimed at investigating the properties of supercontinua generated in tapered fibers pumped at Ytterbium wavelengths around 1 μm, such as noise or the influence of fiber pig-tails and a pulse pre-chirp, that were up to now only known for Ti:sapphire wavelengths around 800 nm [3, 62].

6.2. Properties of Yb:glass as laser medium

Ytterbium doped glasses are very well suited laser materials to build efficient, cost-effective, directly diode-pumped, and compact femtosecond lasers [58, 59]. Besides the advantages that characterize all Ytterbium doped laser materials, for instance a very low quantum defect, a distinct absorption peak at wavelengths of widely available high-power laser diodes, relatively broad emission spectra, and low additional loss processes, Ytterbium glasses additionally show a smoother emission spectrum than most crystals and the manufacturing of laser glasses is usually cheaper than

6.2. Properties of Yb:glass as laser medium

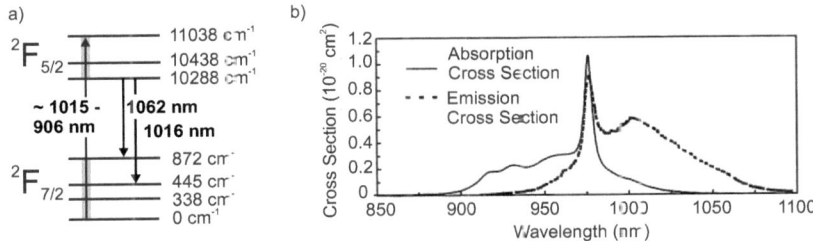

Figure 6.1.: a) Stark levels of the $^2F_{7/2}$ and $^2F_{5/2}$ multiplets of an Ytterbium doped alkali-barium-silicate glass according to [63]. b) Absorption and emission cross sections of 15wt.% Ytterbium doped QX phosphate glass [64].

that of crystals. [58]. Furthermore, glass is an isotropic material and therefore all parameters are polarization independent.

As host material, mostly silicate or phosphate glasses are used. By adding Yb_2O_3 to the molten glass, trivalent Yb^{3+} ions can be inserted into the glass network. [65, 54, 66]. With an electronic configuration of the Yb^{3+} ions of $5s^25p^64f^{13}$, the laser active electrons are located in the incompletely filled 4f shells. Such a configuration keeps the electronic structure simple and is the reason for the low loss processes of Ytterbium doped materials, such as excited state absorption or up-conversion. Furthermore, the electrons in the 4f shells are screened by the outer 5s and 5p shells, and therefore the energy levels are nearly independent on the host material. The only energy levels relevant for laser transitions are corresponding to the $^2F_{7/2}$ and $^2F_{5/2}$ states of the ions. For rare earth as well as lighter ions, this notation of electronic states is usually given in the scheme of L-S-coupling of electron momenta. The laser transitions in all Yb doped laser materials take place between the $^2F_{7/2}$ ground state and the $^2F_{5/2}$ excited state manifold (Fig. 6.1). Nevertheless, despite the screening of the crystal field by outer electrons, the energy levels of the $^2F_{7/2}$ and $^2F_{5/2}$ states split into manifolds because small disturbances due to the electric field of the surrounding host atoms lead to the formation of Stark level multiplets. These disturbances slightly vary from ion to ion and give rise to a host material specific inhomogeneous broadening of the emission spectrum. Since the energy levels involved in pump light absorption as well as stimulated emission belong to the same two Stark multiplets, Yb doped materials are called quasi-three-level materials. At room temperature, the $^2F_{7/2}$ multiplet is thermally populated and the lower laser level is not completely empty. For typical Ytterbium doped laser materials, approximately 4-6% of the

Table 6.1.: Important properties of Ytterbium doped QX laser glass [65, 67].

Laser peak wavelength	λ_L	1032 nm
Absorption peak wavelength	λ_P	976 nm
Max. emission cross section	σ_{em}	$1.4 \cdot 10^{-20}$ cm^2
Max. absorption cross section	σ_{abs}	$0.7 \cdot 10^{-20}$ cm^2
Fluorescence lifetime	μ	2 ms
Index of refraction @ 1032 nm	n	1.52
Thermal conductivity	λ	$0.85 \, \frac{W}{mK}$
Thermo-optical coefficient (20-40° C)	$\frac{\partial n}{\partial T}$	$-21 \cdot 10^{-7}$ K^{-1}
Nonlinear index	n$_2$	$3.36 \cdot 10^{-16} \, \frac{cm^2}{W}$
Group velocity dispersion @ 1045 nm	D$_2$	$+45 \, \frac{fs^2}{mm}$

states of the lower laser level are populated at room temperature and a certain threshold of absorbed pump light is required to achieve transparency.

In nearly all Ytterbium doped materials, the absorption spectrum shows several broad peaks in the region around 910-950 nm as well as a strong narrow maximum around ~980 nm. Laser emission is possible in a range between 1020 nm to more than 1060 nm. The emission spectra are usually sufficiently broad to generate femtosecond laser pulses, for example pulses even shorter than 60 fs were obtained from Yb:glass lasers [58]. Fig. 6.1 representatively shows the involved electronic states of an Ytterbium doped alkali-barium-silicate glass, as well as the emission and absorption spectra of the specific Yb:glass (Yb/QX glass, Kigre Inc.) that was used in this work. The most prominent absorption peak of the Yb doped QX laser glass appears at 976 nm, whereas laser emission is possible in a range between 1020-1060 nm.

As laser material in this work, pieces of a 9.5wt.% Ytterbium doped phosphate laser glass were used as laser medium (Yb/QX glass, Kigre Inc.). We chose this glass because of its high thermal loading limit of more than 11 W per mm absorption length [65], which can be attributed to the low thermal expansion coefficient [68] and makes the laser medium very robust. The most important properties of this laser glass are summarized in table 6.1. The group velocity dispersion was estimated by the Sellmeier equation for a similar phosphate glass (QE-7, Kigre Inc.) [1] [67].

---

[1] $n(\lambda) = 1.5291 + 4433 \cdot 10^{-18} m^2/\lambda^2$.

6.3. Experimental setup of the Yb:glass oscillator

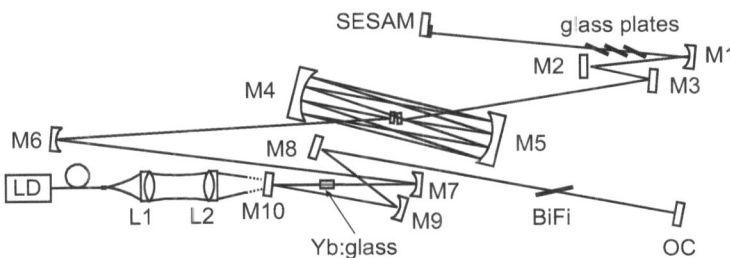

Figure 6.2.: Schematic of the laser oscillator setup: LD pump laser diode; L1,L2 achromatic lenses; M1, M4-7, M9 curved mirrors; M3,M10 dichroic plane mirrors; M2,M8 dispersive plane mirrors; OC 5% output coupler; BiFi birefringent filter; SESAM saturable absorber mirror. The beam path inside the multi-pass cell is only indicated for clarity. In fact the beam bounces nine times off each of the mirrors M4 and M5.

6.3. Experimental setup of the Yb:glass oscillator

6.3.1. Laser medium and pump configuration

The experimental setup of the Yb:glass laser oscillator is schematically shown in Fig. 6.2. As laser medium, a piece of 9.5wt.% Ytterbium doped phosphate glass (Yb/-QX glass, Kigre Inc.) was used. Increasing the doping level to 15wt.% Ytterbium lead to an increase of the laser output power by approximately 20%, however, the laser medium can easily get damaged when the laser operation is started or interrupted too abruptly. When using a maximum doping level of 9.5wt.%, no damage of the laser medium occurred, even if the pump light was turned on and off repeatedly for many times.

The dimensions of the laser medium were $5 \times 2 \times 4$ mm^3 with an optical path length inside the gain medium of 4 mm at normally incident laser beam. In the non-lasing case, approximately 70% of the pump light was absorbed for a pump wavelength of 976 nm. The laser glass was mounted between two aluminum plates. To improve the thermal contacts between laser medium and heat sink, the gaps between the laser glass and the mount were filled by 100 μm thick layers of indium foil. The heat flow was mostly uni-directional along the side with 2 mm height. The aluminum plates were cooled by a TEC and no further heat removal by water cooling was necessary for stable laser operation. In order to minimize losses and to prevent spurious reflections, the laser glass was cut with a small wedge and both front surfaces were antireflection coated for the pump wavelength around 976 nm as well as for the laser wavelength around 1045 nm.

Due to the polarization insensitivity of the laser medium, a fiber coupled laser diode module emitting unpolarized light was employed as pump source. This allowed to efficiently match the laser mode with simple pump optics consisting of only two lenses (Fig. 6.2). The pump source was a fiber coupled diode module (UM5200, Jenoptik Unique Mode GmbH) which provided up to 5.2 W pump power at 976 nm from a multimode fiber with 50 µm core diameter and a NA of 0.15. The beam quality of this laser diode module was approximately 12 times diffraction limited, resulting in a M^2-factor of $M^2 \approx 12$ [69]. The pump beam was collimated by an achromatic lens L1 with a focal length of $f = 30$ mm and re-focused into the laser medium with an achromatic lens L2 with a focal length of $f = 75$ mm. After being injected into the laser resonator through a dichroic mirror (M10), the pump beam formed a spot with a mode radius of 63 µm in the laser medium. This estimation is based on ray-transfer matrix calculations [48] with the appropriate M^2-factor taken into account. The thermal lens induced by the absorption of pump light and heating of the laser crystal was determined by comparing experimentally obtained stability limits of a simple four-mirror resonator with ray-transfer matrix calculations. The focal length of the thermal lens was estimated to $f_{thermal} \approx 20$ mm for the highest pump power of 5.2 W.

6.3.2. Layout of the laser resonator

The laser resonator consisted of a z-folded cavity with a resonator length of 7.5 m which leads to a repetition rate of 20 MHz. The laser medium was placed slightly asymmetrically between two curved mirrors M7 and M9. Both mirrors had a focal length of 75 mm each and focused the laser beam into the laser medium. The laser resonator was extended by a Herriott-type multi-pass cell (MPC) (M4, M5) and two curved mirrors M1 and M6 with a focal length of 250 mm and 1000 mm, respectively. Mirror M1 focused the laser onto the saturable absorber mirror (SESAM). Mirror M3 was a flat dichroic mirror to filter out remaining pump light and prevent it form saturating the SESAM. Two flat dispersive mirrors M2 and M8, as well as several glass plates, were used to adjust the intra-cavity dispersion and SPM in order to compensate for the dispersion of the laser medium and the effects of self-phase modulation. The net intra-cavity GDD per round trip was adjusted in such a way to provide sufficient anomalous GDD to operate the laser in the solitary mode-locking regime [34, 35]. As output coupler (OC), a flat mirror with a reflectivity of 95% was

6.3. Experimental setup of the Yb:glass oscillator

Figure 6.3.: Photo of the setup of the Yb:glass laser oscillator. The SESAM and the birefringent filter were removed at the time the photo was taken. The beam path in the laser resonator is depicted by the yellow lines, whereas the orange shaded area visualizes the pump beam. OC: output coupling mirror; Yb:glass: laser medium; SESAM: position of the saturable absorber mirror, N-SF57: glass plate for dispersion adjustment.

used. The footprint of the whole laser was fit onto an area of 62×23 cm^2 (Fig. 6.3) by folding more than 70% of the total resonator length to a distance of only 30 cm by employing a Herriott-type multi-pass cell [70].

The expected beam waists at distinct positions in the laser resonator were first estimated by simple ray-matrix calculations to approximately 53 μm in the laser medium, and to 220 μm on the SESAM (Fig. 6.4). These calculations were performed by two methods with equal results, either by neglecting the MPC and modeling the resonator by an effective shorter resonator or by taking the MPC into account (Fig. 6.4) (see section 6.3.3). In mode-locked operation, the intra-cavity peak intensities increase tremendously, so that nonlinear effects in the optical elements influence the resonator properties and cannot be neglected anymore. By modeling the laser resonator with the extended ray-matrix method as described in Chapter 5.2, the beam waist inside the laser medium and on the SESAM can be obtained in dependence on the laser output power (Fig. 6.5). Furthermore, the simulations give an estimate for the expected pulse width.

As input parameters for the simulations, we used the properties of Yb:glass (table 6.1) as well as the laser parameters described above. For experimentally achievable laser output powers between 650-790 mW, the beam waist inside the laser medium increases, whereas the beam waist on the SESAM decreases compared to the values for cw operation (Fig. 6.5). The case of cw operation is equivalent to a very low output power in the diagrams of Fig. 6.5. The increase inside the laser medium helps to match the laser mode with the top-hat shaped pump beam of the fiber-coupled diode module with an calculated beam waist of 63 μm. Devia-

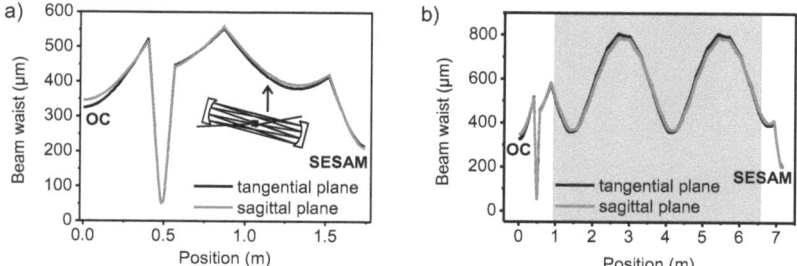

Figure 6.4.: Laser mode radius at each position in the laser resonator calculated with the ray-transfer matrix method without taking nonlinear effects into account. OC and SESAM indicate the two resonator end-mirrors. a) Equivalent short resonator without MPC. The arrow marks the position at which the Herriott-type MPC was inserted. b) Resonator with MPC. The shaded region indicates the region of one round-trip in the MPC.

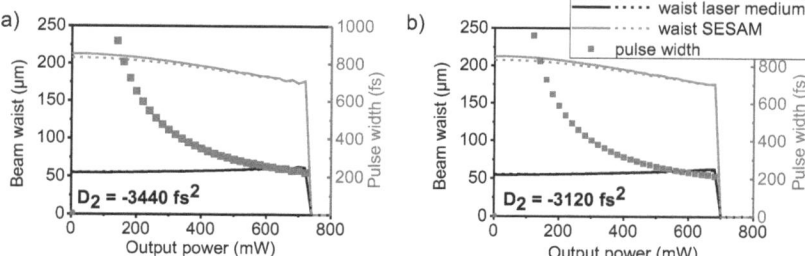

Figure 6.5.: Calculation of the beam waists inside the laser medium and on the SESAM (solid lines: tangential plane, dashed lines: sagittal plane) in dependence on the laser output power. The squares denote the expected pulse width. The resonator was modeled by an extended ray-transfer matrix method to take the optical nonlinearities of the laser medium into account. a) Results for an inserted GDD of D_2 = -3440 fs^2. b) Values obtained for a GDD of D_2 = -3120 fs^2.

6.3. Experimental setup of the Yb:glass oscillator

tions compared to the experimentally obtained values arise one the one hand due to general limitations of the rather simple numerical model which are described in section 5.2. On the other hand, additional discrepancies appear because the non-linearities in the glass plates as well as in the birefringent filter were not taken into account.

6.3.3. Design of the Herriott-type multi-pass cell

A Herriott-type multi-pass cell (MPC) is an arrangement of two mirrors which are aligned in such a way that they form a stable laser resonator by themselves. A laser beam could be infinitely long trapped inside such a cell. However, by inserting a small injection and a pick-up mirror into the cavity of the MPC, the laser beam can be in- and output coupled. Since a Herriott-type MPC forms a stable resonator, the complex Gaussian q parameter is preserved after each round-trip. A round-trip comprises not necessarily twice the distance between the two mirrors, but rather a certain integer number of whole forth- and back-reflections. This number n is specific for each multi-pass cell and is a design parameter. In the following, in contrast to the nomenclature common for laser resonators, a whole propagation through a MPC will be called a round-trip, whereas two consecutive reflections on the two MPC mirrors will be referred to as a forth- and back-reflection.

As long as a laser beam is picked up from a Herriott MPC after exactly an integer number of round-trips, such a multi-pass cell can be inserted into any stable laser resonator at any position without changing any of its beam properties, except the total length. Since a Herriott-type MPC does not affect a laser resonator, the design of the final oscillator can be started with a short cavity. Fig. 6.4 shows the calculated beam waist at each position of the laser resonator. Fig 6.4 a) shows the results for the equivalent laser resonator without MPC. This calculation was used to meet all other design criteria such as the right beam waist inside the laser medium as well as the beam diameter on the SESAM. In the experimental setup, the MPC was then inserted at a suitable position, which is indicated in the diagram by the arrow. Fig 6.4 b) shows the same calculation for the resonator with inserted MPC and the full length. The shaded area visualizes the distance over which the laser beam propagates inside the MPC. One recognizes that in all regions except the MPC, both resonators behave identical. The beam waist inside the MPC changes periodically, but the beam waists and divergence angles at the entrance and exit of the cell are identical.

In order to find a specific combination of curved mirrors and the appropriate mirror distance, beam propagation inside a Herriott-type MPC can be described by ray-transfer matrices [70, 71]. Due to the q preserving behavior, the components of the ray-matrix for one forth- and back-reflection M_T need to fulfill the same condition as for stable laser resonators [48]:

$$\left| \frac{A+D}{2} \right| \leq 1, \qquad (6.1)$$

where A and D are components of the overall ray-transfer matrix of the multi-pass cell $M_T = \begin{pmatrix} A & B \\ C & D \end{pmatrix}$. By interpreting Eq. (6.1) in terms of an angle θ,

$$\frac{A+D}{2} = \cos\theta, \qquad (6.2)$$

the overall matrix after n forth- and back-reflections can be written the following form [71]:

$$M_T^n = \begin{pmatrix} \frac{A-D}{2} \frac{\sin n\theta}{\sin\theta} + \cos n\theta & B \cdot \frac{\sin n\theta}{\sin\theta} \\ C \cdot \frac{\sin n\theta}{\sin\theta} & \frac{D-A}{2} \frac{\sin n\theta}{\sin\theta} + \cos n\theta \end{pmatrix}. \qquad (6.3)$$

It can be shown that for one round-trip inside such a MPC, the beam path in general lies on a hyperboloid [70]. Successive bounces after each forth- and back-reflection between the mirrors form elliptical or circular patterns on the mirrors. The angle θ is exactly the angular separation of two successive bounces.

To be able to preserve the complex beam parameter q after n forth- and back-reflections, hence one round-trip, the matrix M_T^n needs to be equal to the unity matrix I, i.e. $M_T^n = \pm I$. This if fulfilled if θ is a rational fraction of π [71]:

$$\theta = m/n \cdot \pi, \qquad (6.4)$$

where n is the number of bounces on one mirror and m is a parameter which determines the order in which the spot pattern on the mirrors is formed.

In this experiment, we decided to use exactly one round-trip in a Herriott-cell with the parameter $m = 2$ and $n = 9$ bounces per mirror and round-trip. For this combination of m and n, it is possible to achieve a beam path inside the multi-pass cell which lies on the surface of a cylinder. The nine spots on each mirror form a circular pattern in this case and only the first and the last spot overlap, so it is possible to extract the beam after exactly one round-trip. Furthermore, we decided for a symmetric setup with two identically curved mirrors with radius R. The mirrors for beam injection and extraction were placed in the middle of the two curved mirrors (inset of Fig. 6.4). The overall transfer-matrix M_{MPC} for such a cell can be derived by multiplying the ray-matrices for a curved mirror with radius R,

6.3. Experimental setup of the Yb:glass oscillator

$M_M = \begin{pmatrix} 1 & 0 \\ -\frac{2}{R} & 1 \end{pmatrix}$, and the ray-matrix for free beam propagation along a distance L_0, $M_P = \begin{pmatrix} 1 & L_0 \\ 0 & 1 \end{pmatrix}$:

$$M_{MPC} = \begin{pmatrix} 1 - \frac{6L_0}{R} + \frac{4L_0^2}{R^2} & 2L_0 - \frac{2L_0^2}{R} \\ \frac{4L_0}{R^2} - \frac{4}{R} & 1 - \frac{2L_0}{R} \end{pmatrix}. \tag{6.5}$$

The length L_0 of such a multi-pass cell can thus be calculated as a function of the mirror radius of curvature R, the number of bounces n, and the parameter m:

$$L_0 = R \cdot \left(1 \pm \sqrt{1 - \frac{1 - \cos\left(\frac{m}{n}\pi\right)}{2}}\right). \tag{6.6}$$

For the solution with the negative root in Eq. (6.6), the radius of curvature is large compared to the distance between the mirrors. This leads to a beam path inside the multi-pass cell which lies on the surface of a cylinder. For the solution of Eq. (6.6) with the positive root, the beams would intersect at the center.

In the experiment, the Herriott-cell consisted of two mirrors with a diameter of 25.4 mm and a radius of curvature of 5000 mm (Layertec GmbH, Germany). The reflectivity of each mirror was 99.95% with the coating centered on a wavelength of 1040 nm. The multi-pass cell was aligned for nine bounces on each mirror ($m = 2$, $n = 9$), resulting in additional losses of less than 2% per one round-trip in the whole laser resonator. The necessary mirror distance of the multi-pass cell was derived by Eq. (6.6) to $L_0 = 301.5$ mm, resulting in an overall path length of $L_{total} = 5427$ mm for one round-trip. As injection and extraction mirror, two small trapezoidal flat mirrors were used, that were inserted at exactly one segment of the circular cross section of the beam path.

6.3.4. Solitary mode-locking and SESAM

In order to improve the laser stability and provide a rugged solution, we decided to operate the laser in the solitary mode-locking regime [34, 35], i.e. mode-locking is initiated and stabilized by a saturable absorber mirror (SESAM). The laser design can fully focus on stability and the laser resonator can be operated in the middle of a stability regime. To provide the conditions for solitary mode-locking, sufficient net anomalous GDD per round-trip was introduced into the laser resonator. By employing two dispersive mirrors (M2 and M8, Layertec GmbH, Germany) as dispersive elements, the laser resonator can be kept compact. Each reflection on one of these mirrors provided a GDD of -900 fs$^2 \pm 100$ fs^2. The GDD as well as the SPM per round trip could additionally be fine-tuned by inserting uncoated plates of N-SF57 glass into the beam path at Brewster angle. Each of the 1 mm thick plates provided a GDD of approximately +80 fs^2 per single pass. Furthermore, one end-mirror was replaced by a SESAM [36, 37, 39]. The SESAM had a modulation depth of $\Delta R = 1.6\%$ and a saturation fluence of 70 μJ/cm^2 (Batop GmbH, Germany). The laser mode was focused onto the SESAM by the curved mirror M1 with a focal length of 250 mm to a spot size of 440 μm in diameter for cw-operation and to an estimated diameter of 350 μm, respectively, when the influence of the Kerr-lens is considered. According to the limit derived in [42], the above mentioned parameters would lead to the onset of Q-switching for modulation depths larger than $\Delta R < 0.9\%$. However, the value for the saturation fluence of the SESAMs was taken from the datasheets of the manufacturer which were not necessarily correct for the specific manufacturing runs of the SESAMs. The experiments showed, that a modulation depth of $\Delta R = 1.6\%$ lead to stable and self-starting mode-locked operation. For lower modulation depths, mode-locking was either not self-starting or did not start at all. Even with the SESAM strongly saturated with a fluence of 690-820 μJ/cm^2, no double-pulsing or cw-breakthroughs were observed.

An increase in stability and the possibility to tune the laser wavelength could be achieved by inserting a 1 mm thick crystalline quartz plate into the laser resonator at Brewster angle. This quartz plate acted as a birefringent filter [72]. For the laser configuration with 180 fs pulse-width, the center wavelength was tunable between 1038 nm and 1047 nm. For the configuration with 150 fs pulses, for stable mode-locked operation the laser wavelength had to be tuned to a suitable regime in the emission spectrum. A center wavelength of either 1040 nm or 1045 nm resulted in stable cw mode-locking.

6.4. Characterization of the laser parameters

Fig. 6.6 a) shows a measurement of the obtained laser output power in dependence on the pump power. During this measurement, the laser was configured to provide a pulse width of 147 fs at the maximum pump power. Mode-locking started at a pump power of approximately 4.5 W, as indicated by the step of the output power, as well as the increase of the slope efficiency from $\eta = 17.5\%$ to $\eta = 19\%$. The laser was self-starting in the mode-locked regime and showed pure single-pulse mode-locking for pump powers above 4.5 W. For pump powers below 4.5 W the laser operated in the cw regime and no tendency toward Q-switching could be observed. The maximum achievable output power for this configuration was 665 mW at a pulse width of 147 fs and was only limited by the available pump power. Fig. 6.7 a) shows autocorrelation

6.4. Characterization of the laser parameters

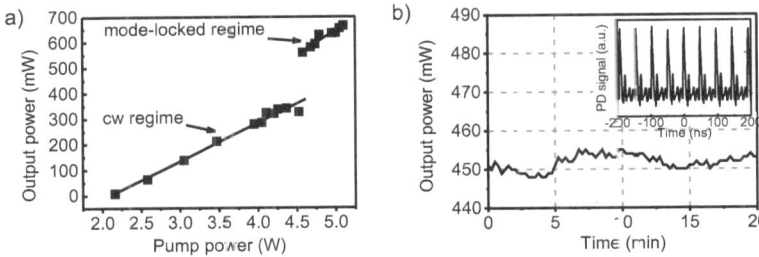

Figure 6.6.: a) Measurement of the laser output power in dependence on the pump power for a laser configuration with an inserted GDD of $D_2 = -3120$ fs^2. Mode-locking started at a pump power of approximately 4.5 W. The solid lines depict a linear fit for the two different operation regimes. b) Output power at a pulse width of 147 fs over a span of 20 min. The intensity noise over this period was determined to be 0.43% of the rms intensity. The inset shows a photo-diode signal of the 20 MHz pulse train.

traces for this case. Assuming a perfectly sech^2-shaped pulse envelope, the FWHM of the pulses was determined to be 147 fs. The corresponding optical spectrum for this case showed a bandwidth of approximately 8 nm (Fig. 3.7 b) at a center wavelength of 1044 nm, corresponding to a nearly transform-limited time-bandwidth product of 0.32. In the mode-locked regime, a stable 20 MHz pulse train could be observed by measuring the laser intensity with a fast photo-diode (inset Fig. 6.6 b). The laser output power was stable and mode-locking was not interrupted for many hours. Fig. 6.6 shows a measurement of the laser output power for a pulse width of 147 fs over a span of 20 min. During this measurement not the full laser power was coupled into the power meter. From this measurement, the intensity noise was determined to be 0.43 % of the rms power over 20 min.

The pulse width could be adjusted by changing the net GDD per round-trip by varying the number of glass plates that were inserted into the laser resonator. The overall inserted GDD was then determined by the combination of dispersive mirrors and glass plates. By inserting three glass plates, corresponding to an inserted GDD of $D_2 = -3120$ fs^2, the pulse width of 147 fs at an maximum average power of 665 mW was obtained. With only one glass plate, corresponding to an inserted GDD of $D_2 = -3440$ fs^2, pulses with 180 fs pulse width at an maximum average power of 790 mW were obtained with an optical-to-optical efficiency of 15%. In

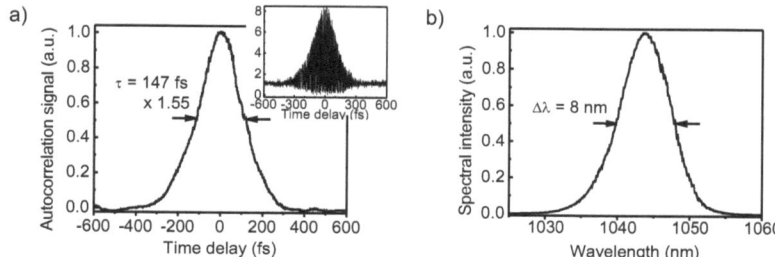

Figure 6.7.: a) Intensity autocorrelation trace for a laser configuration with an inserted GDD of -3120 fs² per round-trip. The pulse width was determined to be 147 fs, assuming a perfectly sech²-shaped pulse envelope. The inset shows the trace of an interferometric autocorrelation. b) The laser spectrum showed an bandwidth of approximately 8 nm at a center wavelength around 1044 nm.

both cases, the pulses were transform-limited with a time-bandwidth product of 0.32. The interferometric autocorrelation (inset Fig. 6.7 a) further indicates that the pulses were nearly transform-limited and unchirped.

The theoretical predictions for the pulse widths (Fig. 6.5) vary significantly from the experimental results. The main reason for this is that for the simulations with the extended ray-transfer matrix method, all influences were neglected that were due to nonlinear effects in the glass plates as well as in the birefringent filter. However, each glass plate induces a significant amount of self-phase modulation and a Kerr-lens is induced in the plates. Besides all general problems of this numerical model, neglecting of the self-phase modulation in the glass plates yields an explanation for the significantly shorter pulses that were obtained experimentally. For example in the case of three inserted glass plates and thus an inserted GDD of $D_2 = -3120$ fs², the additional SPM in 3 mm of N-SF57 glass leads to a pulse width of 147 fs compared to a pulse width of ∼200 fs as is predicted theoretically for this case. Furthermore, neglecting the Kerr-lenses that is induced in the plates would additionally change the dependence of the beam waists on the output power (Fig. 6.5). Experimentally no limit was observed for output powers of 700 mW or 740 mW, respectively (Fig. 6.5).

Table 6.2.: Order N of initially launched solitons in tapered fibers with different waist diameters D. As input parameters $\lambda = 1040$ nm, $P_0 = 600$ mW, $\eta = 50\%$, $\tau_{FWHM} = 160$ fs, $f_{rep} = 20$ MHz were used.

D (μm)	GVD (ps/nm·km)	β_2 (fs²/mm)	N
2.0	+175 [30]	-1004	6
2.7	+90 [30, 13]	-516	6
4.3	+25 [13]	-143	8

6.5. Supercontinuum generation with the Yb:glass laser

6.5.1. Measurements with tapered fibers

For peak intensities sufficiently high so that supercontinuum generation is significantly affected by the fission of higher order solitons, propagation lengths in microstructured fibers of only a few cm can lead to the generation of multi-octave broad spectra. In this case, most likely tapered fibers are one of the most flexible and a very cost-effective method. Tapered fibers can be manufactured within minutes and without complex equipment. This allows to easily vary parameters such as waist diameter, taper length, or even taper shape.

The order of the initially launched solitons that can be achieved with this setup can be calculated by the relation resulting from soltion theory [28]

$$N^2 = \frac{\gamma \hat{P} \tau}{|\beta_2|}, \qquad (6.7)$$

with the SPM parameter γ, the peak power $\hat{P}$, the paramter τ related to the FWHM pulse width τ_{FWHM} by $\tau_{FWHM} = 1.76 \cdot \tau$, and the fiber GVD parameter β_2. This equation can further be expressed by the experimentally obtained parameters for the input power P_0, coupling efficiency η, repetition rate f_{rep}, and fiber diameter D:

$$N^2 \approx \frac{2.58 \cdot n_2 \cdot P_0 \cdot \eta \cdot \tau_{FWHM}}{\lambda \cdot D^2 \cdot f_{rep} \cdot |\beta_2|} \qquad (6.8)$$

The obtained values are summarized in table 6.2 for three different fibers with waist diameters of 2.0 μm, 2.7 μm, and 4.3 μm, respectively. As nonlinear index n_2, a value of $n_2 = 2.6 \cdot 10^{-20}$ m²/W for SMF28 fiber [28] was used and the input power was set at $P_0 = 600$ mW with a coupling efficiency into the fiber of $\eta = 50\%$. The calculations confirm that this combination of Yb:glass laser and tapered fibers

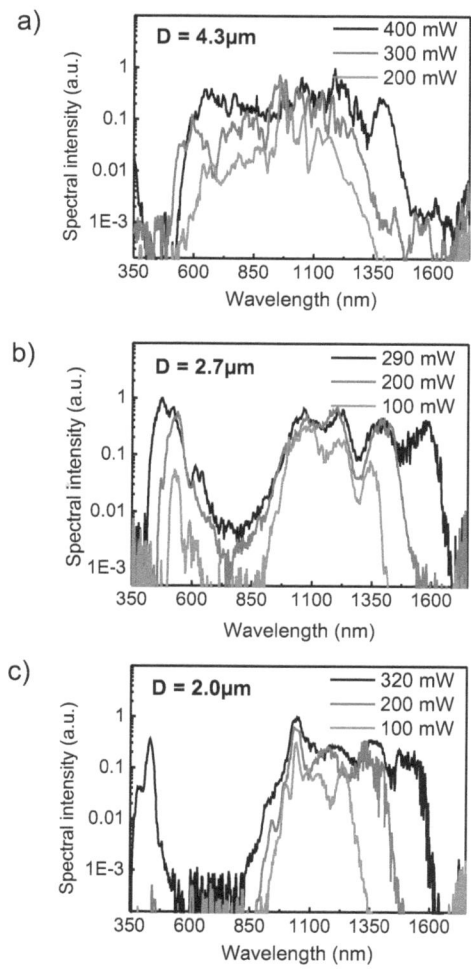

Figure 6.8.: Supercontinua obtained from three different tapered fibers. The measurements are shown for no pulse pre-compression for various average powers of the supercontinua. a) Spectrum with 400 mW average power achieved with a fiber with a diameter of 4.3 μm. b) Broad spectrum obtained at 290 mW average power obtained with 2.7 μm fiber waist diameter. c) Spectrum of a fiber with 2.0 μm waist diameter with a significant fraction in the blue-ultraviolet region at a total average power of 320 mW.

6.5. Supercontinuum generation with the Yb:glass laser

generate solitons of the order of six to eight. This means that the fission of higher order solitons plays an important role and the supercontinuum generation process can be described by the theory developed for femtosecond pulses [3].

In order to demonstrate possible spectra that can be obtained with such a combination of the Yb:glass oscillator and tapered fibers, Fig. 6.8 shows power dependent measurements of supercontinua generated with three different tapered fibers. During all measurements, the laser was operated with two intra-cavity glass plates, resulting in a pulse width of 160 fs and a laser power of approximately 700 mW. To protect the laser against back-reflections from the fiber surface, a Faraday isolator (EOT, Inc.) was placed between the laser output coupler and the fiber coupling optics. The laser light was then coupled into the tapered fiber by a 10× microscope objective. Behind the pump optics, approximately 600 mW average power remained. The input power into the fibers could additionally be attenuated by a reflecting filter wheel. All fibers were pig-tailed with approximately 10-15 cm SMF 28 fiber at both ends. The thin tapered region was 90 mm long for all fibers. The waist diameter of each fiber varied between 2.0 μm and 4.3 μm.

Fig. 6.8 b) shows the experimentally obtained power dependent spectra of the supercontinua generated with a tapered fiber with 2.7 μm waist diameter. With an input power of 600 mW, an overall supercontinuum average power of 290 mW was achieved. The spectrum stretched from 400 nm to 1650 nm. Less broad but smoother spectra were obtained by employing a tapered fiber with a thicker taper waist of 4.3 μm (Fig. 6.8 a). The maximum average output power was 400 mW in this case. Spectral components reaching into the violet to near ultraviolet region were obtained by using fibers with small diameters. For instance, with a waist diameter of 2.0 μm, a spectrum with up to 320 mW average power and a significant fraction of light in the blue to ultraviolet region was achieved (Fig. 6.8 c).

In all three diagrams of Fig. 6.8 one can nicely see the influence of the waist diameter on the shape of the spectra. All spectra show a strong peak at the visible regime. This peak is due to the fission of higher order solitons during the first stages of supercontinuum generation along propagation over the first 0.5-1 cm in the tapered region [3]. The position at which the non-solitonic peak in the visible appears, is dependent on the phase matching condition (Eq. 3.2) which is strongly dependent on the waist diameter, as well as on the input power. For thinner fibers, the distance between pump wavelength and zero-dispersion wavelength is larger and thus the visible peak appears at shorter wavelengths than in thicker fibers. Furthermore, all spectra show a broad and smooth fraction of light at longer wavelengths

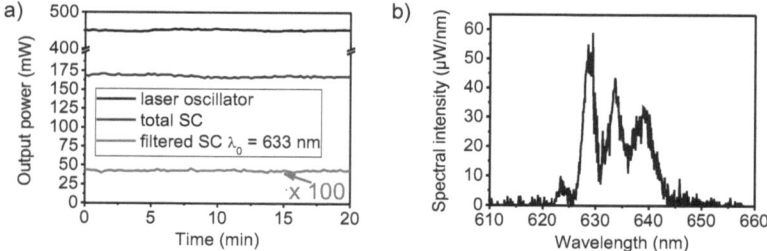

Figure 6.9.: a) Measurement of the laser output power, the overall average power of a supercontinuum generated by a fiber with 4.0 µm waist diameter, as well as the average power of a 15 nm broad spectral section centered around 633 nm. b) Spectrum of the supercontinuum filtered by a bandpass filter with a center wavelength of 633 nm and a bandwidth of 15 nm.

than the pump wavelength. These regions mostly stretches from the pump wavelength around 1044 nm to the near-infrared. These long-wavelength components are partially generated during the fission of higher order solitons, but mostly and to a large extend due to the self-frequency shift of fundamental solitons caused by Raman scattering [3]. The longer the interaction length in a fiber is, the further the near-infrared limit is shifted toward longer wavelengths. The gap between the peak at visible wavelengths and the red-shifted components is successively filled by four-wave mixing over the whole propagation length. However, the larger the distance between the visible peak and the infrared section, the more distinct is the gap in the spectrum. For large distances (e.g. Fig. 6.8 b) and c)), this gap usually remains after propagation over typical lengths of ∼10 cm.

6.5.2. Noise measurements

For many applications of supercontinua it is important to maintain a stable spectral power over the duration of the experiment. An important property of filtered supercontinua is an increasing intensity noise of spectral sections with narrower filter bandwidth. In the following, this aspect was investigated for an arbitrary filter center wavelength of 633 nm. In order to characterize the intensity stability and the noise of the generated supercontinua, the overall average power as well as the power of a spectrally filtered narrow section were measured and compared to the laser stability. Fig. 6.9 a) shows the measurement of the laser output power, the overall average power of the supercontinuum, and the average power of the spec-

6.5. Supercontinuum generation with the Yb:glass laser	53

trally filtered section over a span of 20 minutes. During all measurements, the laser was configured with two inserted glass plates and a pulse width of 160 fs. The supercontinuum was generated in a tapered fiber with a waist diameter of 4.0 μm. The spectrally narrow section was filtered out of the supercontinuum by a interference filter with a center wavelength of 633 nm and a bandwidth of 15 nm. The spectrum of the filtered section with an average power of 430 μW is shown in Fig. 6.9 b). Over a span of 20 min, the intensity noise of the overall average power of the supercontinuum was determined to be 0.79% rms which is approximately twice the laser noise of 0.43%. Narrow filtering of supercontinua tremendously increases the noise due to fluctuations of the shape of the spectrum. For exammple, the average power of the 15 nm broad spectrally filtered section showed an intensity noise of 2.3% rms.

6.5.3. Influence of pulse chirp and fiber pig-tails

For practical applications it is important to know the influence of the SMF 28 fiber pigtails that are connected to the tapered fibers, as well as the consequences of a possible chirp of the input pulses. During every-day operation of the fibers, especially under harsh and unprotected conditions, the fiber ends need to be re-cleaved from time to time. Thus, the fiber pigtails will become shorter over time. Besides that, maybe it would be beneficial to exchange a tapered fiber, either to change the shape of the spectrum or to exchange a damaged fiber. Another important factor is the influence of the pulse chirp on the supercontinuum generation. Since the laser beam might pass several optical elements before it is coupled into the fiber, such as a Faraday isolator or the microscope objective, the spectra of the supercontinua also depend on the position of the tapered fiber in the experimental setup. These points turned out to be extremely sensitive for supercontinuum generation with Ti:sapphire lasers [62]. Assuming that self-phase modulation in the pig-tails is negligible, the only noticeable effect of the pig-tails to the laser pulses is a chirp. Therefore, investigating the influence of a pulse chirp on the supercontinuum generation also gives insight on the influence of the fiber pig-tails. The measurements with this particular laser can be regarded as an example for other Ytterbium doped lasers with similar properties such as similar wavelength and peak powers.

In order to investigate the aforementioned effects, the supercontinua of two different tapered fibers were measured for varying negative pulse pre-chirp. The chirp of the input laser pulses was changed by bouncing the laser beam off dispersive mirrors before coupling into the fiber. One bounce on a dispersive mirror provided

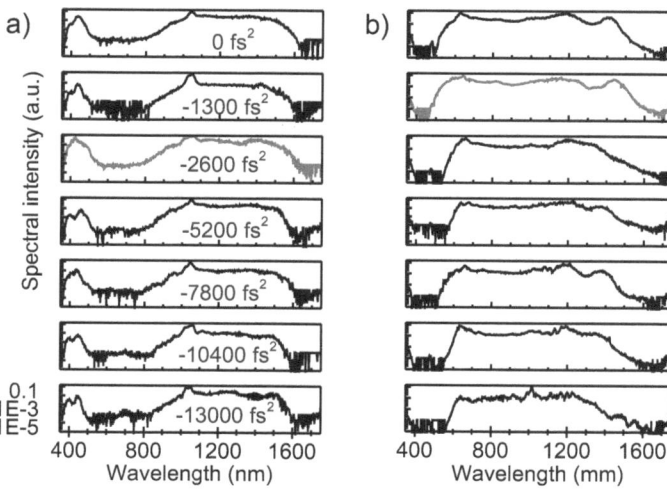

Figure 6.10.: Supercontinua in dependence on negative pulse pre-chirp. The broadest spectra are plotted in red. a) Spectra for a fiber with 2.0 µm waist diameter. The output power was held constant in all cases at 280 mW. b) Same measurement for a fiber with 4.3 µm waist diameter at a constant output power of 325 mW.

-1300±150 fs^2 of anomalous GDD. The number of bounces on the dispersive mirrors was varied between none and ten, resulting in a GDD of 0 fs^2 to -13000 fs^2. Before coupling into the tapered fiber, the laser beam passed a Faraday isolator, the microscope objective, and a 14 cm long fiber pigtail. All these elements accounted for an additional normal pulse chirp. Fig. 6.10 a) shows the measured supercontinua for a fiber with 2.0 µm waist diameter. Fig. 6.10 b) shows the same experiment for a fiber with a diameter of 4.3 µm. The output powers were held constant for each measurement at 280 mW and 325 mW, respectively.

The measurements lead to two conclusions. First, the broadest spectra were obtained for pulses with slightly anomalous chirp. For the fiber with 2.0 µm diameter (Fig. 6.10 a), the broadest spectra were obtained with a pre-chirp of -2600 fs^2, whereas for the fiber with 4.3 µm (Fig. 6.10 b) a pre-chirp of -1300 fs^2 gave the best results. These low values for the optimum pre-chirp are rather obvious, because the pre-chirp helps to compensate the positive GDD of the microscope objective

6.5. Supercontinuum generation with the Yb:glass laser   55

and input fiber pig-tail. When entering the tapered region, normally chirped pulses are initially temporally compressed and experience self-phase modulation until they finally form a soliton. Excess power forms a pulse which temporally broadens and forms a background signal. Consequently, a laser pulse that enters the tapered region with a flat phase forms a soliton immediately and will thus experience a longer effective interaction length than a pulse with an unsuitable phase. The second observation is, that for the fiber with 2.0 μm waist diameter the shape of the spectra varies less with different pre-chirps than the spectra of the fiber with 4.3 μm waist diameter. The reason for this could be that the effects of a pre-chirp are less noticeable for the thinner fiber with a GVD of -1004 fs^2/mm (table 6.2) which by itself would induce the same chirp over a propagation length of 13 mm. The thicker fiber with 4.3 μm waist diameter would induce this chirp over a much longer length of more than 1 m. This fact does not affect the fraction of light which propagates through the fiber as a soliton, but the fraction of excess light which remains and forms the background signal. However, in general one recognizes that a moderate pulse pre-chirp of less than -13000 fs^2 has only little influence on the smoothness and bandwidth of the spectra. This agrees with the results found theoretically as well as experimentally for pumping with Ti:sapphire lasers [62]. On one hand, tapered fibers with typical waist diameters of 1-4 μm show stronger anomalous dispersion at the the taper waist for pump wavelengths around 1045 nm than they do for pump wavelengths around 800 nm. On the other hand, the normal dispersion in the fiber pig-tails is lower for the Ytterbium wavelengths around 1045 nm than it is for Ti:sapphire lasers at 800 nm because 1045 nm are closer to the zero-dispersion wavelength of SMF28 fiber at 1300 nm.

Therefore, fiber pig-tails that are connected to the tapered fibers or a pulse chirp do not have a huge influence on the spectral shape of supercontinua generated with Ytterbium doped lasers. Concerning the shape of the supercontinua, no additional measures need to be taken in order to deal with fiber pig-tails or pulse chirp. This means, that fibers can be cleaved or exchanged, or the position of the tapered fiber in the experimental setup can be changed without a noticeable influence on the generated supercontinua. Besides that, practically no temporal pulse pre-compression is necessary since the effect is hardly noticeable.

6.5.4. Measurements with photonic crystal fiber (PCF)

For femtosecond supercontinuum generation, tapered fibers offer a very flexible solution. Alternatively, nonlinear PCF are widely commercially available. Possibilities

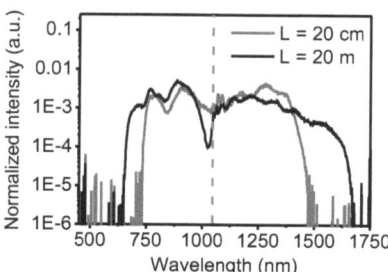

Figure 6.11.: Spectra of two supercontinua obtained by pumping 20 cm and 20 m long pieces of nonlinear PCF with 190 fs pulses at a wavelength of 1040 nm (dashed line). With an input power $P_{in} = 600$ mW, an overall average output power of up to $P_{out} = 215$ mW was obtained.

of pumping nonlinear PCF with the previously described Yb:glass laser will be described in the following section. For the particular fiber used in these experiments, the pump wavelength was very close to the zero-dispersion wavelength, resulting in differently shaped spectra compared to supercontinua obtained with tapered fibers. Latter were pumped far in the anomalous dispersion regime.

For the experiment, the laser was operated an average power of 625 mW, a pulse width of 190 fs and a center wavelength of 1040 nm. The laser was protected against back-reflections by a Faraday isolator and the laser light was coupled into the PCF by a 30 × microscope objective. As micro-structured fiber, a nonlinear PCF with a core field diameter of 5 µm and a zero-dispersion wavelength of $\lambda_{ZDW} = 1040$ nm was used (Thorlabs, SC-5.0-1040). The coupling efficiency into the PCF was approximately 35%, resulting in an average power of the supercontinua of 225 mW. By using different microscope objectives the coupling efficiency was increased, however, the stability of the fiber coupling decreased with tighter focusing.

Fig. 6.11 shows the resulting spectra for 20 m and 20 cm long pieces of PCF, respectively. The spectra differ slightly from the spectra that were generated in tapered fibers in the previous sections. In both cases, the spectra obtained with this PCF are rather narrow and posses little spectral components at wavelengths shorter than the pump wavelength. The longer fiber generated broader spectra due to the longer interaction length and thus higher possibilities for soliton self-frequency shifts and further nonlinear processes such as four-wave mixing and cross-phase modulation.

For the employed PCF, the zero-dispersion wavelength was very close to the pump wavelength of 1040 nm. For such a situation, the blue-shifted wavelength components that evolve due to fission of higher order solitons are in close proximity to the pump wavelength as predicted by the phase-matching condition Eq. (3.2). Furthermore, it might be that some fraction of the pump light experiences normal dispersion and does not form a soliton. This case is comparable to pumping the fiber with a lower input power which leads to less broad spectra, as can be seen for example in the power dependent measurements with tapered fibers (Fig. 6.8). The fraction of pump light which experiences normal dispersion is solely broadened by self-phase modulation. However, newly generated red-shifted components might then again experience anomalous dispersion and contribute to the formation of solitons. In the measurements shown in Fig. 6.11 the peaks due to fission of higher order solitons appears at approximately 875 nm and 750 nm. The longer the fiber, i.e. the longer the interaction length, the soliton self-frequency shift is more pronounced and the spectrum becomes broader toward the long wavelength side. Besides this, Raman scattering and four-wave mixing processes are also more efficient in longer fibers and lead to a further broadening of the spectra toward shorter wavelengths.

Concerning practical applications, rather narrow spectra, such as obtained in this experiment, may be interesting if high spectral intensities are required. For an efficient supercontinuum generation with sub 200 fs pulses at peak powers of 160 kW, fiber lengths of 10-20 cm are sufficient.

6.6. Conclusion

We presented an approach for a flexible and portable femtosecond supercontinuum source for various applications. In order to achieve this, a novel and very compact design for an Yb:glass laser oscillator was developed. The combination of this laser together with tapered fibers can generate supercontinua with up to 400 mW overall average power and various differently shaped spectra. When pumping tapered fibers at wavelengths of Ytterbium lasers, the influence of fiber pig-tails and pulse chirp on the shape of supercontinuum spectra was found to be very small. This behavior could be attributed to the GVD of SMF28 fiber and the waist region of tapered fibers at wavelengths of Ytterbium lasers around 1 μm. Spectrally narrow filtering of the supercontinua with a bandwidth of 15 nm resulted in an intensity noise of less than 2.5% of the rms power in this spectral section. The experimental results of the supercontinuum generation experiments agree well with the theory of femtosecond supercontinuum generation with micro-structured fibers [1, 2, 3] and can thus be well predicted by these theories.

Chapter 7

Multi-Watt 44 MHz femtosecond supercontinuum source based on an Yb:KGW oscillator

In order to scale the average power of femtosecond supercontinua to more than 1 W, a novel design for a diode-pumped multi-Watt, sub-200 fs laser oscillator was developed, and methods for sufficient longitudinal pumping of the gain medium with broad-area diodes were investigated. Pumped by a single broad-area laser diode, the oscillator can provide over 5 W output power and pulse widths as short as 161 fs at a repetition rate of 44 MHz. An optical-to-optical efficiency of more than 23% for a pulse width of 161 fs, and an efficiency of 28% for 225 fs pulses were obtained. We experimentally and numerically investigated the thermal effects that are induced in the slab laser crystal by pumping with up to 18 W from a broad-area diode. We found that the specific beam profile of these diodes has a significant influence on the thermal effects. This profile results in a different behavior than, for instance, fiber coupled diodes or diodes with top-hat beam profiles. Experimental guidelines and a numerical model are presented to predict and compensate these effects by a suitable resonator design. In combination with tapered fibers, supercontinuum generation with an overall average power of more than 1.3 W was achieved. Furthermore, an experiment with polarization-maintaining PCF is presented in order to demonstrate the possibility of generating polarization stable femtosecond supercontinua.

7.1. Current state of research on multi-Watt sub-200 fs laser oscillators

A major challenge in increasing the output power of diode-pumped femtosecond laser oscillators is posed by the therefore required high pump powers. Due to such high pump powers, thermal effects are induced in the gain medium. In recent years, especially Ytterbium doped tungstates proved to be very suitable gain media for diode-pumped femtosecond lasers. Mostly all recent publications on high-power diode-pumped femtosecond lasers are based on those gain materials, see for example Refs. [21, 73, 20, 74]. Such lasers are usually longitudinally pumped by high-power laser diodes. This leads to the evolution of a non-uniform heat distribution inside the gain medium. This temperature distribution changes the index of refraction and causes mechanical strains and stresses. Both effects lead to a change of the optical properties of the gain medium. Some of the effects can be compensated by a suitable resonator design, whereas higher-order aberrations often lead to a low output beam quality. In particular for mode-locked lasers it is important to minimize higher-order aberrations in order to achieve a nearly diffraction limited beam.

In order to overcome spatial variations of the optical path length in the gain medium due to a non-uniform heat distribution, various laser designs have been pursued. Very high average powers and extremely high pulse energies directly from a femtosecond oscillator can be achieved with the concept of thin-disk lasers [20, 74]. In this design, the heat is transferred collinearly with the propagating laser beam. Another possibility is the so-called zig-zag geometry which is mostly used with slab lasers. Here, due to an averaging effect, all fractions of the phase front experience approximately the same phase shift while the beam propagates through the gain material [75]. Especially for high-power femtosecond lasers, the thin-disk concept would be very interesting since the propagation length inside the gain medium with a typical thicknesses of $\sim$250 μm is short which helps to keep distorting nonlinear and thermal effects low. However, there seems to be a limit of approximately 200 fs for pulses that can be obtained from thin-disk lasers [20]. This limit is probably due to a narrowing of the emission spectrum (gain narrowing) caused by the extremely high pump light densities necessary for these kind of lasers. This is for example noticeable by the appearance of cw-backgrounds in mode-locked operation.

Methods to achieve sub-200 fs pulses at multi-Watt output powers are still a current topic of research. Recently, there have been great advances based on the bulk-laser concept with thin, but rather long laser crystal slabs. For example, Holtom obtained more than 5 W average power at 134 fs and 9.9 W at 292 fs from an Yb:KGW oscillator [21]. A very simple design without the need for dichroic mirrors or complex coatings has been demonstrated by Berger et al. [73]. However, the pulse width was limited to 250 fs. Here we demonstrate that further progress is possible with the concept of slab laser oscillators by employing broad-area diodes as pump sources. Furthermore, we show that the N_g-cut crystal orientation, seldomly used up to now, is favorable for such kind of lasers. Due to the extremely high beam quality of a novel kind of pump diodes, as well as thorough considerations for minimizing or compensating for thermal effects, the overall efficiency of slab lasers can nearly achieve values of thin-disk lasers, and yet allow for the generation of sub-200 fs pulses. For example in this work, we achieved an optical-to-optical efficiency of more than 23% at a pulse width of 161 fs and an average power of more than 4 W.

7.2. Ytterbium doped double tungstates

Even though Yb:glass offers many advantages such as isotropy or a broad and smooth emission spectrum, the thermal conductivity of glasses and the fracture limit is too low to increase the laser power to several Watts with a gain medium of only several mm absorption length. Very promising gain media for high-power diode-pumped femtosecond lasers are provided by Ytterbium doped monoclinic double tungstates. These crystals belong to a class of crystals with the general formula Yb:Kx(WO$_4$)$_2$, where x can be one of several possible lanthanide ions such as Y,Gd, or Lu of which a certain fraction is substituted by Yb ions. The most commonly used host materials of this class are KGd(WO$_4$)$_2$ (KGW), KY(WO$_4$)$_2$ (KYW), or KLu(WO$_4$)$_2$ (KLuW). Even a doping level of 100% Yb is possible, resulting in the stochiometric KYb(WO$_4$)$_2$ (KYbW) crystal. With these tungstate laser materials, tremendous advances in the development of diode-pumped solid state femtosecond lasers were achieved recently. For example as mentioned above, Holtom obtained an average power of 5 W at 134 fs and 9.9 W at 292 fs from an Yb:KGW oscillator [21]. Brunner et al. achieved 22 W average power with a pulse width of 240 fs from an Yb:KYW thin-disk laser [20]. Among the tungstates, Yb:KLuW is the most newly developed material. It is particularly very promising for the epitaxial growth of higly doped thin films [76, 77]. With this material, Rivier et al. could recently demonstrate an Yb:KLuW based thin-disk laser with single-pass pumping [78].

Ytterbium doped tungstates posses some interesting properties and are highly

Table 7.1.: Coefficients of the Sellmeier equation (Eq.(7.1)) to determine the refractive index for each axis N_p, N_m, and N_g, respectively [84].

	A	B	C/nm	D/nm^{-2}
N_p	1.5344	0.4360	186.18	$2.0999\cdot10^{-9}$
N_m	1.5437	0.4541	188.91	$2.1567\cdot10^{-9}$
N_g	1.3867	0.6573	170.02	$0.2913\cdot10^{-9}$

suitable as gain media for high power diode-pumped femtosecond lasers. On one hand, these materials show the same characteristics as all Ytterbium doped materials (see for example the properties of Yb:glass as described in Chapter 6.2). On the other hand, due to their crystalline structure, these materials show reasonably high thermal conductivities that are sufficient to achieve laser powers in the range of several tens of Watts. For example, Yb:KGW has a thermal conductivity of 2.6-3.8 W/m·K [79, 80]. This value ranges between that of laser glasses (0.85-1.30 W/m·K) and highly thermally conductive materials such as YAG with a thermal conductivity of 14 W/m·K [81]. However, despite their crystalline structure, the emission spectrum of Yb tungstates is sufficiently broad to generate laser pulses shorter than 100 fs. For example Liu et al. obtained pulses as short as 71 fs from a Kerr-lens mode-locked Yb:KYW laser [82]. Another property of Yb doped tungstates are their high emission and absorption cross sections of approximately $\sigma_{em} \approx 3 \cdot 10^{-20}$cm^2 and $\sigma_{em} \approx 12 \cdot 10^{-20}$cm^2, respectively [83]. These high cross sections allow building very efficient lasers and help to suppress Q-switching in mode-locked operation with a SESAM.

7.3. Femtosecond 5 W Yb:KGW slab oscillator

7.3.1. Properties of Yb:KGW

In this work, we decided to use Yb:KGW as gain medium because of the greater availability as well as the slightly smoother emission spectrum for a laser polarization parallel to the N_m refractive axis as compared to Yb:KYW. It was first employed as gain material for mode-locked lasers by Brunner et al. who could generate an average power of 1.1 W at a pulse duration of 176 fs from a bulk laser oscillator that was pumped by two 3 W laser diodes [85].

Yb:KGW is a highly anisotropic crystal. Nearly all properties are dependent on

7.3. Femtosecond 5 W Yb:KGW slab oscillator

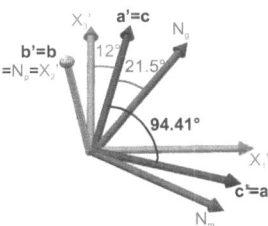

Figure 7.1.: Crystal axes of Yb:KGW: a, b, and c (blue) with respect to the principal refractive axes N_p, N_m, and N_g (red), as well as the principal axes of thermal expansion X_1', X_2', X_3' (green).

the specific crystal orientation. In the literature, some properties, such as emission and absorption cross sections, as well as the thermal conductivity, are often specified in the coordinate system of the crystallographic axes. Other properties such as the refractive indices or the thermal expansion can be more conveniently described in their own principal coordinates. The crystal structure of KGW can be described by different monoclinic space groups and additionally in each group with different unit cells. Most publications and manufacturers use lattice parameters of a unit cell similar to the ones used by Pujol et al. [86]:

$$a' = 7.582 \text{ Å}, \quad b' = 10.374 \text{ Å}, \quad c' = 8.087 \text{ Å}, \quad \beta = 94.41°.$$

These lattice parameters correspond to a description in a I2/a space group. As pointed out by Pujol et al., many publications use similar unit-cell parameters, however, with inappropriate space groups. In particular older publications often use similar parameters as Pujol et al. in combination with a C2/c space group [86]. Furthermore, many authors use similar unit cells with the axes labeled with a, b, and c which would correspond to $a = c'$, $b = b'$, $c = a'$. In the following we will refer to the lattice parameters of Pujol et al. with the labeling of a, b, and c since this is the most widely used combination in literature and datasheets.

Concerning the linear optical properties, KGW is a biaxial birefringent crystal with three main refractive axes. The index of refraction along the principal refractive axes of the indicatrix are labeled by N_p, N_m, N_g, where p, m, and g stand for petit, médian and grand, respectively [87]. All three principal refractive axes are perpendicular to each other. In respect to the crystallographic axes, the N_g-axis is rotated by 21.5° clockwise from the a'-axis, the N_m-axis is rotated by 17.1° clockwise from the c'-axis and the N_p-axis is parallel to the b'-axis [84]. The wavelength

Chapter 7: Multi-Watt 44 MHz femtosecond supercontinuum source

Table 7.2.: Material properties of Yb:KGW. All anisotropic values are given in the according principal system [79, 80, 84, 86, 88].

Parameter			Value (∥ axis)		
			a	b	c
Emission peak wavelength	λ_L	(nm)	1027	1032	1024
Absorption peak wavelength	λ_P	(nm)	981	981	981
			934	953	953
Em. cross section @ λ_L	$\sigma_{em,L}$	(10^{-20} cm^2)	2.8	2.2	0.77
Abs. cross section @ 981 nm	$\sigma_{abs,P1}$	(10^{-20} cm^2)	12.0	1.7	2.1
Abs. cross section @ 934/953 nm	$\sigma_{abs,P2}$	(10^{-20} cm^2)	2.7	1.8	0.7
Thermal conductivity @ 100°C	k	(W/m·K)	2.6	3.8	3.4
			X_3'	X_2'	X_1'
Thermal expansion coefficient	α	(10^{-6}K^{-1})	10.64	2.83	23.44
			N_m	N_p	N_g
n @ 1027 nm			2.011	1.983	2.062
GVD @ 1027 nm	k_2	(fs^2/mm)	+183	+170	+217
Nonlinear index @ 800-1600 nm	n_2	(10^{-16}cm^2/W)	20	15	-
Thermo-optic parameter	$\partial n/\partial T$	(10^{-6} K^{-1})		0.4	

dependence of the refractive indices can be described by a Sellmeier equation [84]:

$$N_{p/m/g} = A + \frac{B}{1 - \left(\frac{C}{\lambda}\right)^2} - D\lambda^2. \tag{7.1}$$

The coefficients A, B, C, and D are listed in table 7.1 for each axis.

Another effect which is often described in its own principal coordinate system is the thermal expansion. The axes of the principal system of thermal expansion are labeled as X'_1, X'_2, and X'_3. The X'_2-axis is parallel to the b'-axis, whereas the X'_3-axis is rotated by 12° anti-clockwise from the c'-axis and is perpendicular to the X'_1-axis [86].

Fig. 7.1 visualizes the orientation of the crystallographic as well as the principal axes of the indicatrix and the axes of thermal expansion. The most important material parameters of Yb:KGW are summarized in table 7.2. Due to the anisotropy

7.3. Femtosecond 5 W Yb:KGW slab oscillator

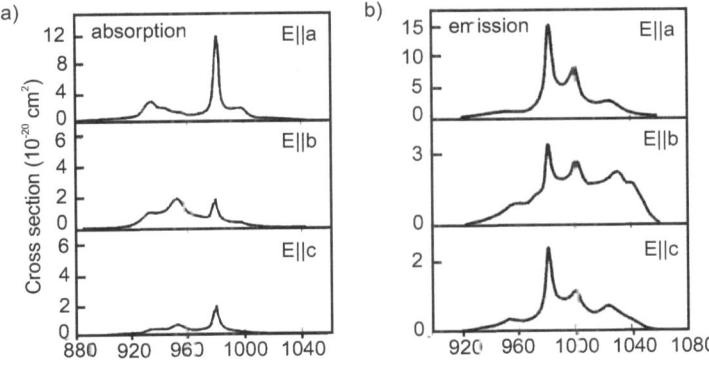

Figure 7.2.: Absorption (a) and emission cross sections (b) of 5at% doped Yb:KGW [79].

of the crystal, the parameter values are given along each principal axis. If necessary, the tensor of these parameters can be converted to other coordinate systems. Some parameters, such as the thermo-optical parameter $\partial n/\partial T$, vary significantly in different references. Here, the values which to our experience correspond best with the experimental results are given. When important for the experiment or the simulations of the thermal lens, the specific parameters are discussed in detail in the corresponding chapters.

Fig. 7.2 shows the absorption and emission cross sections of 5at.% Ytterbium doped KGW [79]. The spectroscopic properties of Yb:KGW are similar to other Yb doped materials (see Chapter 6.2). The laser transitions take place between the $^2F_{5/2}$ excited state and the $^2F_{7/2}$ ground state manifolds. The most prominent absorption peak for Yb:KGW appears at exactly 981 nm with a very narrow linewidth of only 3.7 nm. A second but broader absorption peak appears depending on the crystal orientation at 934 nm or 953 nm, respectively. Laser emission is possible in a range between 1023-1060 nm, whereas the emission peak wavelength λ_L is strongly dependent on the direction of polarization. Also the smoothness of the emission spectrum is dependent on the light polarization and crystal orientation.

7.3.2. Broad-area laser diodes

In recent years, end-pumped bulk femtosecond lasers have made significant advances. These advances are partially due to new laser host materials such as Yb doped tungstates, but also to a large extend to the availability of high quality pump diodes

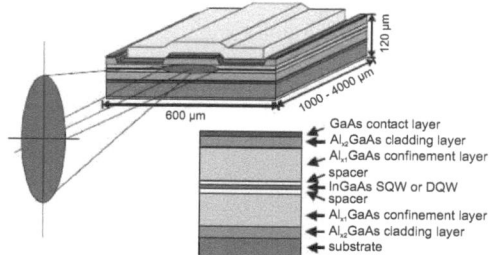

Figure 7.3.: Typical layout and layer structure of a broad-area laser diode [24].

which allow for efficient pumping of those gain materials [85, 21, 22].

One major advantage of end-pumped solid-state lasers is that the pump beam and the laser beam can completely overlap in the gain medium and no excess pump power is wasted. Thus, the efficiency of end-pumped solid state lasers can be higher compared to other laser types, for example flash-light or side-pumped lasers. Furthermore, the heat load that is deposited in the laser medium can be kept at a minimum. A sufficient condition for the case that both beams completely overlap in the gain medium is given if both beams have a beam waist of approximately the same size in the gain medium. In most cases, this is limited by the pump beam due to its bad beam quality compared to the laser mode.

In particular for quasi-three level materials such as Ytterbium doped gain media, the intensities need to be high to achieve transparency and efficient laser operation. For common laser diodes with several tens of Watts of pump power, this implies pump beam diameters in the range of 10-1000 μm. Additionally for mode-locked lasers, high intensities help to suppress an unwanted modulation of the pulse train due to Q-switching.

For laser diode bars or stacks where the output of several emitters is combined, typical M^2-factors are on the order of $M^2 \approx 1 \times 1000$. With such values, one is limited to rather short laser crystals or one has to use large pump spots. Short laser crystals on one hand require a strong absorption in order to maintain high efficiencies which can lead to strong thermal effects. Large pump spots on the other hand reduce the laser efficiency. This is especially noticeable for Yb doped gain media with high pump thresholds necessary to even achieve transparency. Advances are possible by balancing the M^2-value equally on both beam axes with the help of beam-shaping optics like prisms or small lenses. With such beam-shaped diodes, M^2-values on the

7.3. Femtosecond 5 W Yb:KGW slab oscillator

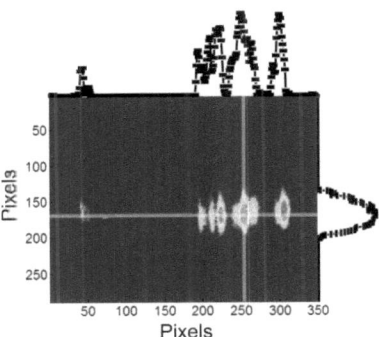

Figure 7.4.: CCD-camera image of the beam profile of a 12 W broad-area diode. The cross sections show the normalized intensity distribution along the slow- and the fast-axis, respectively.

order of $M^2 \approx 50 \times 50$ can be achieved.

A further progress in simplicity and efficiency of end-pumped solid-state lasers is possible by employing so called broad-area diodes [24, 25, 26, 89]. These laser diodes can emit up to 47 W in cw-operation from a single emitter with a nearly diffraction limited beam along the fast axis [89]. Along the slow axes M^2-factors of $M^2 \approx 50 - 60$ can be achieved for powers up to 18 W as with the diodes used in this work. These diodes are based on MQW-structures as shown in Fig. 7.3. For wavelengths around 980 nm, InGaAs is used as material for the MQW and GaAs and AlGaAs for confinement-, cladding- and contact-layers. The facets are usually coated with a dielectric Bragg structure. One facet provides a reflectivity of approximately 95%, whereas the other facet acts as ouptut coupling mirror with a reflectivity around 10% [24]. The tremendous progress in high power laser diode technology can be mainly attributed to two major developments. On one hand, a novel and very thoroughly optimized design and structure of the epitaxial layers as well as the choice of dopants reduces the ohmic series resistance and the free-carrier absorption. At the same time, this novel design provides a waveguide structure which is known as the large optical cavity (LOC) concept. Here, one uses a several hundred μm broad and at the same time a rather thick waveguide, with thicknesses around 2-3 μm. The thereby reduced modal gain is compensated by employing very long cavities with lengths of more than 1 mm [24]. The LOC concept decreases the intensities on the diode facets and additionally keeps the far-field divergence angle on the slow-axis low. Even waveguides broader than 3 μm can be used in the so-called SLOC-concept.

Chapter 7: Multi-Watt 44 MHz femtosecond supercontinuum source

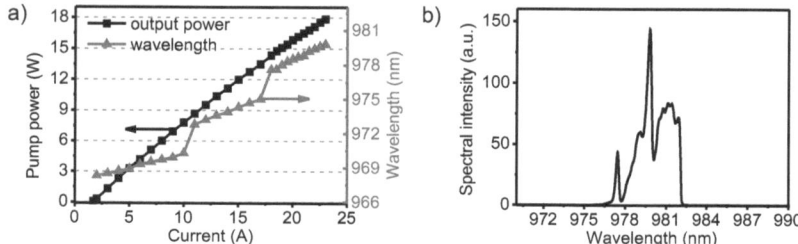

Figure 7.5.: a) Output power (black squares) and center wavelength (red triangles) of an 18 W broad-area emitter at a constant temperature of $T = 25°C$. Both the shape of the spectrum and the center wavelength change with increasing current. The reason for the jumps of the plotted center wavelength is a change of the shape of the spectrum and a shift of the position of the pre-dominant peak which was considered as center wavelength. b) Spectrum at $T = 25°C$ for an output power of 18 W. Here the distinct peak was centered at 979.8 nm.

Here, a mode selection is necessary, which is possible since modes with a node in the active area experience less gain and are discriminated against other modes. The second major improvement which enabled the development of high-power broad-area diodes is due to the tremendous progress in facet coating technology. The key point is to provide very clean coatings on the facets, because even tiny absorption of light leads to a vicious cycle of band-gap narrowing and more absorption, which finally ends in a so-called catastrophic optical mirror damage (COMD) [89]. In fact, in case of broad-area emitters, COMD is not the limiting factor for the output power but a thermal roll-off, i.e. an decrease of the optical power due to non-radiative losses. Further progress can be attributed to improved manufacturing techniques and extremely high purities of the semiconductor materials and dielectrics that are used, as well as improvements in wafer handling and manufacturing processes, mainly MOCVD or MBE.

The typical beam profile of such a broad-area diode is that of an approximately Gaussian-distribution of a diffraction-limited beam along the fast axis. Along the slow axis, the beam profile is a pattern with several peaks that are formed due to a superposition of many different transversal modes. Fig. 7.4 shows an image of the beam profile of a 12 W broad-area emitter that was taken with a CCD-camera. The cross sections show a nearly Gaussian distribution along one axis and a complex pattern along the other axis.

7.3. Femtosecond 5 W Yb:KGW slab oscillator

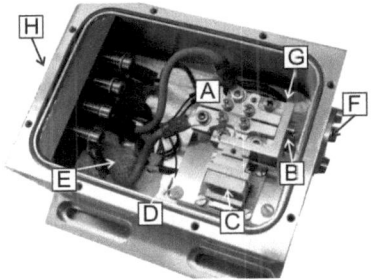

Figure 7.6.: The broad-area diodes were mounted in a sealed box. A: broad-area diode in CCP-package, B: aspherical lens, C: xyz-translation stage, D: TEC and Ni-coated copper plate, E: silica-gel, F: AR-coated laser window, G: thermistor, H: feed-throughs for the electrical and thermistor connections as well as connections for water-cooling.

In this work, we used two different types of broad-area emitters, that were provided by the Ferdinand-Braun-Institut für Höchstfrequenztechnik, Berlin. The diodes with the chip mounted on a CCP-package were capable to emit 12 W or 18 W, respectively from a 200 μm wide emitter with a waveguide thickness of approximately 3 μm. The spectral width of the diodes was 3 nm and ~4 nm, respectively (Fig. 7.5 b). All diodes were designed to operate at a wavelength of 981 nm at the maximum current and temperatures of 20-25°C. Fig. 7.5 a) shows the output power and center wavelength of an 18 W broad-area diode in dependence on the current for a constant temperature of $T = 25°C$ measured at the position of the temperature sensor. Fig. 7.5 b) shows the corresponding spectrum at maximum current of $I = 23$ A. The beam quality of these diodes is nearly diffraction limited with approximately $M^2 \approx 1.2$ along the fast axis, and still with a very good beam quality of $M^2 \approx 50\text{-}60$ along the slow axis.

The diodes were mounted in a sealed box in order to protect them from dust and humidity (Fig. 7.6). Inside the box, the diode was mounted on a Ni-coated copper-plate which could be cooled by a TEC. In order to remove the heat, the whole box could be water-cooled. The temperature of the plate in proximity to the diode was measured with a NTC thermistor. In order to collimate the light along the fast axis, an aspherical lens with a focal length of $f = 3.1$ mm was placed in front of the diode. The lens could be adjusted by a xyz-translation stage. After assembling of the box, a pack of silica gel was inserted inside the box to reduce the humidity.

Chapter 7: Multi-Watt 44 MHz femtosecond supercontinuum source

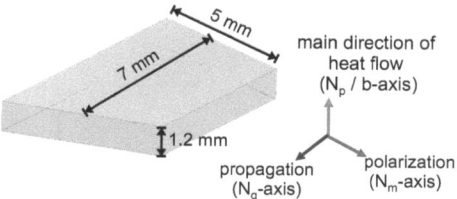

Figure 7.7.: Schematic of the employed flat-Brewster cut Yb:KGW crystal slab. The coordinate system depicts the orientation of the laser polarization, propagation direction, and the major direction of heat flow in respect to the crystal geometry.

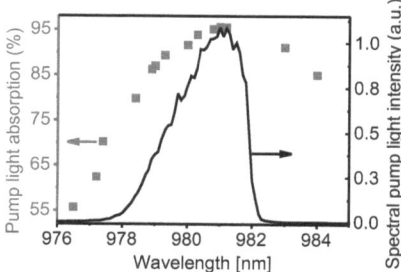

Figure 7.8.: Wavelength dependence of the pump light absorption in the Yb:KGW crystal for the non-lasing case (red squares). The black line indicates the spectral intensity of a 12 W broad-area diode.

7.3.3. Geometry, orientation and properties of the Yb:KGW slab

The described high-power laser oscillator in this work employs a slab laser crystal design. By using a very thin but long gain medium, the heat can be removed along the thin side and the laser light can propagate through the gain medium along the long path length. In this way, problems due to too high inversion can be avoided and sub-200 fs pulses are feasible.

Fig. 7.7 schematically shows the geometry of the employed Yb:KGW crystal. In order to fully exploit the good beam quality of the broad-area diodes along the fast and the slow axis, we decided to use an elliptical beam inside the laser crystal with an ellipticity of ∼2. With such an elliptical beam, it is feasible to obtain the smallest possible beam diameters for which the confocal lengths along the fast and the slow axis can be kept similar to the crystal length. To more easily match the laser mode with such an elliptical pump mode, an end-pumped resonator design with a flat Brewster-cut crystal was used. This concept proved to show good results for laser crystals longitudinally pumped by laser diodes, see for example refs. [90, 91, 36]. The flat side of our laser crystal was AR-coated for the pump-wavelength around 981 nm, as well as for the laser wavelength around 1027 nm. As laser medium we employed a N_g-cut oriented Yb:KGW crystal, with the laser- and pump-light polarization both parallel to the N_m-axis. The absorption length in the crystal was 7 mm at a doping level of 1at.% Ytterbium. Measurements showed, that approximately 95% pump-light at 981 nm were absorbed in the non-lasing case (Fig. 7.8). The crystal was 5 mm wide and 1.2 mm high. The heat was mostly extracted along the side of 1.2 mm height, and thus parallel to the N_p-axis for which KGW shows the highest thermal conductivity. This specific combination of crystal geometry and orientation was chosen because it allows using a flat Brewster-cut crystal and yet keeping the laser beam propagation direction parallel to the optical table, whereas the heat can be extracted in the perpendicular direction. However, in this case one has to take into account that the emission spectrum for the laser polarization along the N_m-axis is less smooth than for the more commonly used polarization along the N_p-axis. Due to the long absorption length of 7 mm, the thermal load can be kept low and the thermal lens can be very well treated as parabolic lens in proximity of the laser mode, as discussed in more detail in the next section.

7.3.4. Experimental characterization of the thermal effects and cw-operation of the Yb:KGW oscillator

To achieve stable cw-mode-locking, it is necessary to provide a laser resonator which generates a nearly diffraction limited laser mode. With up to 18 W of pump light absorbed in a few mm long crystal, as well as with peak powers on the order of MW in mode-locked operation, both thermal and nonlinear optical effects in the laser crystal, significantly influence the properties of laser oscillators. Therefore, thermal effects and nonlinear optical effects need to be taken into account in the laser resonator design and it is necessary to assure that these effects can be compensated in the first place.

Before investigating nonlinear optical effects which will only be noticeable in mode-locked operation, the influence of thermal effects were studied experimentally as well as with a simple numerical model (section 7.3.5). As discussed in

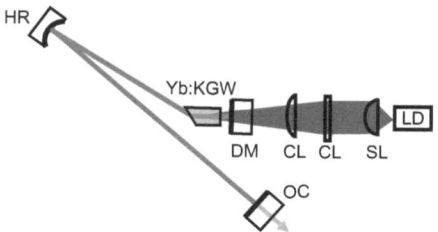

Figure 7.9.: The focal length of the thermal lens as well as the properties in cw-operation were determined by investigating simple three-mirror resonators. The focal length of the thermal lens was obtained by measuring the divergence angle of the laser mode at the output coupling mirror. LD: laser diode, SL and CL: one aspherical and two cylindrical lenses, respectively, HR: curved highly reflective mirror, DM: dichroic mirror, OC: output coupling mirror.

Chapter 5.1, the thermal effects induced in an end-pumped gain medium that have a direct influence on the laser mode. They can be described as a combination of an ideal lens, the so-called thermal lens, plus additional higher order optical aberrations. In most cases, the ideal thermal lens can be easily compensated by the resonator design as long as the focal length is not too short compared to the focal lengths of the other optical components used in the resonator. Higher order optical aberrations, however, distort the phase front of the laser mode in a more complex and possibly random way and can not be compensated by simple measures. Such aberrations would require a precise knowledge of the exact influence on the phase front and a custom compensation for every laser configuration, or an averaging of the phase front like in the zig-zag laser design [75]. Hence, for practical longitudinally end-pumped femtosecond laser oscillators, it is of crucial importance to keep higher order aberrations negligible.

In order to measure the focal length of the induced thermal lens and to test if higher order aberrations on the laser mode destroy the phase front in such a way that it would prevent mode-locking, various simple three-mirror resonators were build (Fig. 7.9). As high-power broad-area emitters, a 12 W and a 18 W broad-area emitter at 981 nm center wavelength were used. The pump beam with a beam quality of $M^2 \approx 1.2 \times 50\text{-}60$ was emitted from a 200 μm broad emitter with a waveguide thickness of approximately 3 μm. The output beam was first collimated with an aspherical lens with a focal length of $f = 3.1$ mm and then refocused through a dichroic mirror into the laser medium with two cylindrical lenses with $f = 200$ mm

7.3. Femtosecond 5 W Yb:KGW slab oscillator

and $f = 50$ mm focal length, respectively. The spot size inside the gain medium had a calculated diameter of approximately $D \approx 200 \times 400$ μm^2. The losses from all three AR-coated lenses added up to approximately 9% of the pump light. The laser resonators consisted of a flat dichroic mirror which also acted as one of the resonator end-mirrors. Furthermore, two different curved mirrors with radii of $R = 300$ mm and 200 mm, respectively, were used to focus the laser mode into the gain medium in order to match the laser mode diameter to the pump beam diameter. A second flat partially reflective mirror acted as output coupling mirror. The transmission of the output coupling mirrors was varied between 2.5% and 5%.

Each laser resonator was optimized to provide an output beam with a beam quality of $M^2 < 1.1$. The beam quality was measured with a commercial beam propagation analyzer (Coherent, Modemaster). The optimization of such a laser resonator is an iterative process. The procedure is started with low pump powers for which the thermal lens is still negligible and the pump power is step by step increased during the optimization procedure. Starting with a resonator design that is optimized for a thermal lens with an infinite focal length, the pump power is increased up to a value for which a decay of the beam quality can be observed. At this point, the focal length of the thermal lens is measured, and an adapted resonator is calculated. This procedure has to be repeated until a beam quality of approximately $M^2 = 1$ is achieved for the highest pump power. For such an adapted resonator, the beam quality can be worse for lower pump powers, or the laser threshold can be much higher compared to the starting resonator, because the thermal lens is required to form a stable resonator.

The focal length of the thermal lens was obtained by comparing the experimentally obtained value of the divergence angle of the laser mode at one position with a ray-matrix calculation of the specific resonator. The thermal lens is thereby treated as two simple cylindrical lenses, that are inserted into the center of the gain medium (see Chapter 5.1). We measured the divergence angle of the output beam of the resonator since it can be accessed most easily at this position. The measurements were performed with the same beam propagation analyzer that was also used to determine the beam quality (Coherent, Modemaster). In order to reliably retrieve the focal length of a thermal lens, it is necessary to keep the beam quality as close to $M^2 = 1$ as possible. A value of $M^2 > 1$ can imply both, an increased beam waist or an increased divergence angle compared to the ideal case. In principle, the beam propagation analyzer could also retrieve the beam diameter of the internal waist inside the laser resonator. Since both methods give identical

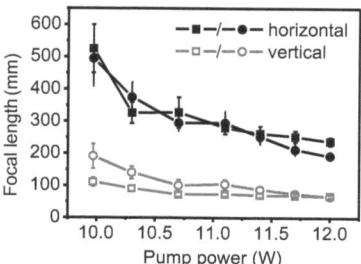

Figure 7.10.: Focal length of the thermal lens along two axes induced by pumping the Yb:KGW crystal with the 12 W broad-area diode. It was determined experimentally by measuring the beam divergence of the output beam of two different three-mirror resonators. (squares: resonator with a curved mirror with 200 mm radius, circles: resonator with a curved mirror with 300 mm radius).

results, we decided for a measurement of divergence angles. Even though there exist somewhat more precise experimental methods, for example based on measuring with an additional probe laser [92] or Z-scan measurements [93], the method of measuring divergence angles allows to conduct the experiment with the final setup and crystal mount and no additional optics need to be inserted. Furthermore, by employing the divergence angle method, the thermal lens can be exactly measured for the laser wavelength, whereas for probe lasers mostly different wavelengths are used.

Fig. 7.10 shows the experimentally obtained focal length of the thermal lens in dependence on the pump power for two different laser resonators that were both pumped with the same 12 W broad-area diode. The black squares depict the results obtained for a resonator with a curved mirror with a radius of $R = 300$ mm, whereas the red circles show the results of a different resonator with a mirror with $R = 200$ mm. As indicated by the error bars, this experimental method exhibits many uncertainties. Here, only the error of the divergence angle measurement and the error of the measurement of the angle of incidence on the curved mirrors were taken into account. Nonetheless, this method is sufficient to find optimized resonators that can provide a beam quality of $M^2 < 1.1$ for maximum pump powers of 12 W and 18 W, respectively (Fig.7.11). The obtained approximate values for the focal length are sufficient for a suitable resonator design, both, for cw as well as for mode-locked operation. Due to the remaining small uncertainties, each resonator usually needs

7.3. Femtosecond 5 W Yb:KGW slab oscillator

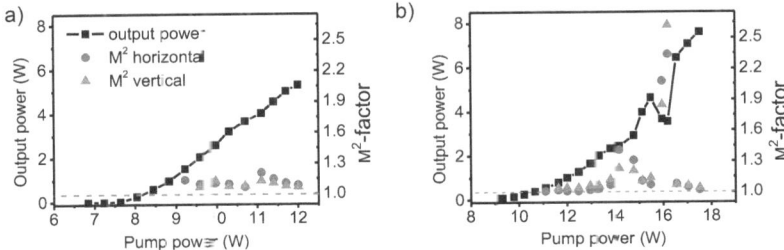

Figure 7.11.: Output power (black curve, squares) and M^2-values (circles: horizontal, triangles: vertical) for a resonator pumped with a 12 W diode (a) and one pumped with a 18 W diode (b), respectively. The laser wavelength was approximately 1030 nm in both cases and the pump wavelength ranged from ∼976 nm to 981 nm for the highest pump power in each case.

to be further empirically optimized in the lab after a simulation. Alternatively, numerical simulations might offer a more accurate prediction of the focal lengths in the future (section 7.3.5).

Fig. 7.11 further shows that, in addition to a good beam quality, the optimized laser resonators show a high efficiency. The maximum optical-to-optical efficiency in cw-operation was $\eta = 44\%$ in the case of pumping with the 12 W diode resulting in an output power of 5.3 W. In the case of pumping with the 18 W diode, an efficiency of $\eta = 43\%$ at a maximum output power of 7.7 W was obtained. In the first experiment, an output coupling transmission of 2.5% was used, whereas in the latter experiment, a 5% output coupler was employed. The beam quality for the resonator pumped by the 18 W diode (Fig. 7.11 b) shows a large jump around 14-16 W pump power. This behavior can either be attributed to a mismatch of the specific thermal lens for these powers and the resonator layout. Another possibility could be that the resonator was aligned in such a way to overlap the laser mode with a peak of the pump beam profile. While the pump power is turned up, the pump beam profile changes. Thus, the overlap between laser mode and peaks of the pump beam profile changes while the pump power is turned up. Also a combination of both explanations is possible. Besides that, the influence of the pump beam profile can also be noticed in other measurements, as discussed later.

As mentioned above, we experimentally observed several effects that suggest a significant influence of the specific beam profile of a broad-area emitter. For instance, the dependence of the M^2-value of the laser mode on the pump power can

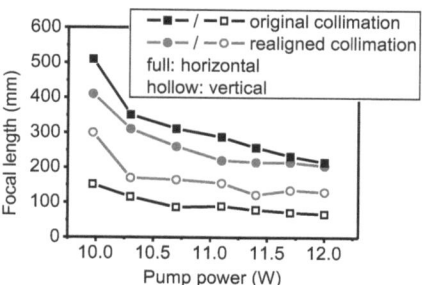

Figure 7.12.: Measurements of the thermal lens with the same 12 W diode. The squares depict the measurements with the original collimation and the circles show the results obtained after a slight re-alignment of the collimation lens.

be described for each section with a smooth behavior, whereas for some pump powers jumps of the M^2-value appear. This behavior is for example very pronounced in the diagram of Fig. 7.11 b) at a pump power of 14 W and for a pump power of 16 W. To a much weaker extend, the same behavior is also recognizable in Fig.7.11 a) for 11 W pump power. Another observation was made, when the aspherical collimation lens with $f = 3.1$ mm focal length in close proximity to the broad-area emitters was slightly re-aligned by a lateral shift of only a few μm. Fig. 7.12 shows the measured focal lengths before and after the re-alignment. Since this lens is very close to the diode, it influences the diode output due to very small back-reflections. If the lens is moved while the diode is emitting light, changes of the beam pattern can be observed. A third observation could be made by exchanging the broad-area diode with a different one and measuring the focal length of the thermal lens in each case. Fig. 7.13 shows the experimentally obtained focal lengths of the thermal lens for three different broad-area diodes. In all three cases, the beam diameters inside the laser crystal were approximately identical. One recognizes, that it is even possible to achieve a stronger thermal lens with 12 W pump power than with 18 W. Employing even a pump power of 18 W and leaving all other parameters the same, a completely different pump power-dependence of the thermal lens can be observed. If the diode was from the same manufacturing run, the deviations between the diodes were less pronounced than for diodes of different manufacturing runs. As supported by numerical simulations in the next section, this behavior is probably due to the resulting beam profile of a specific combination of lenses and diode. Furthermore, the beam profile of one diode is not constant, but changes significantly for different

7.3. Femtosecond 5 W Yb:KGW slab oscillator

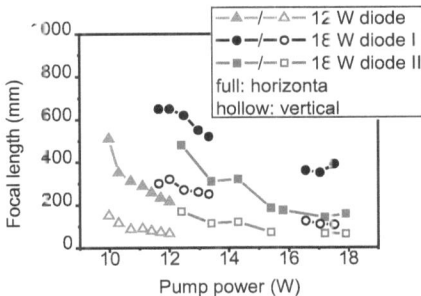

Figure 7.13.: Measurements of the focal length of the thermal lens that were obtained for pumping with three different broad-area diodes (one 12 W diode, two different 18 W diodes). Even if the beam diameters were held the same in all three cases, the thermal lens can be even stronger in the case of *less* pump power.

power settings. Fig. 7.14 shows three images of beam profiles that were taken at different pump powers. The irregular beam-profile along the slow-axis with pronounced peaks, leads to huge deviations from the behavior that were expected from a source with an ideal Gaussian or top-hat beam.

Due to the varying behavior concerning the thermal lens for different broad-area diodes, either each diode needs to be experimentally characterized or, as shown in section 7.3.5, the thermal lens can be to some extend predicted from a measured beam profile. However, due to the big differences between the thermally induced focal lengths for different diodes, the laser resonator needs to be adapted to each diode individually. For lower pump powers, or with further laser diode development in the future, the laser resonator design could be further simplified by the use of tapered laser diodes that emit a beam with an intensity profile that is closer to a Gaussian beam than the irregular beam of broad-area emitters [24, 94].

7.3.5. Numerical modeling of the thermal lens induced by pumping with broad-area diodes

In order to further investigate the experimentally observed influence of the beam profile of broad-area diodes on the thermal lens, the thermal effects in the laser crystal were modeled numerically and compared to experimental results. All numerical simulations described in the following section were performed according to

78 Chapter 7: Multi-Watt 44 MHz femtosecond supercontinuum source

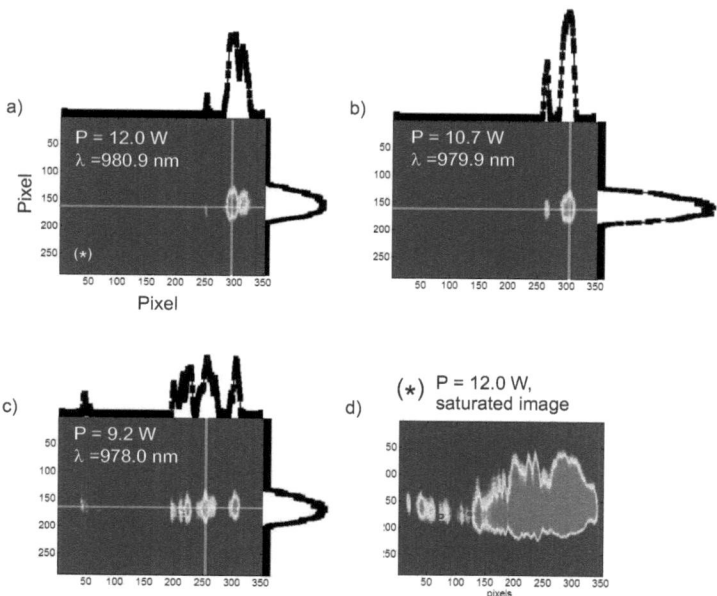

Figure 7.14.: Normalized intensity distribution of beam-profiles of a 12 W broad-area taken with a CCD-camera. The cross section plots visualize the intensity profile along the slow- and the fast-axis. The images were measured after the first aspherical collimation lens. One recognizes that the pattern along the slow-axis varies for different power settings. The saturated picture (d) reveals the full details of the beam profiles that are not visible on a linear scale.

the method described in Chapter 5. All experimental determinations of the focal lengths were done according to the procedure described in Chapter 7.3.4.

Before comparing the numerical model according to Ch. 5 to the experiment, all material parameters need to be identified or estimated. Many properties of Yb:KGW still vary significantly in the literature. This can be due to variations of the properties from crystal to crystal, as well as general reasons such as the direction of crystal growth [97]. Further uncertainties arise due to difficulties to achieve reliable measurements for many parameters. For example, the thermo-optical parameter $\partial n/\partial T$ is particularly vaguely known since it is difficult to distinguish the contribution of the thermo-optical effect from the usually simultaneously occurring photo-elastic

7.3. Femtosecond 5 W Yb:KGW slab oscillator

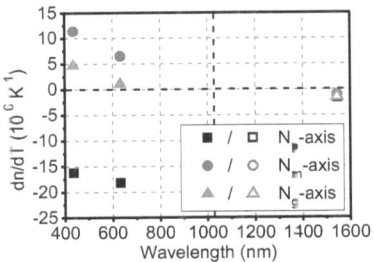

Figure 7.15.: Wavelength dependent values of the thermo-optic parameter $\partial n/\partial T$ for a light polarization along each refractive axis as specified by the two references Filippov et al. [95] (full symbols), and Biswal et al. [96] (hollow symbols). The laser wavelength is indicated by the vertical dashed line.

Table 7.3.: Material parameters that were used for the FEA simulations of the thermal effects in the Yb:KGW crystal.

Parameter		Value (|| axis)			Ref.
		a	b	c	
Thermal cond. @100°C k	(W/mK)	2.6	3.8	3.4	[79, 80]
$\partial n/\partial T$	(10^{-6} K^{-1})		0.4		[79, 88, 95, 96]
$\partial n/\partial T$ (fit)	(10^{-6} K^{-1})	0.2	0.9	-	
Thermal-exp. coeff. $\alpha_{a,b,c}$	(10^{-6} K^{-1})	4	5.6	8.5	[79]
Specific heat C	(J/kg·K)		500		[80, 79]
Young's modulus E	(10^9 Pa)	112.4	152.5	94.2	[80]
Poisson's ratio ν			~0.3		[79]

effect. Furthermore, both effects are wavelength dependent, anisotropic and thus often described in different principal coordinates. For instance, the value for the thermo-optical parameters $\partial n/\partial T$ for a wavelength around 1030 nm and a polarization along the N_m-axis varies between $-1.0 \cdot 10^{-5}$ K^{-1} and $+4.3 \cdot 10^{-6}$ K^{-1} in the literature [80, 98]. In this work, a value close to $\partial n/\partial T = +0.4 \cdot 10^{-6}$ K^{-1} for laser light around 1030 nm and polarized parallel to the N_m-axis was used. This decision was based on the specification of the crystal manufacturer [79], as well as on several references that, to our experience, sufficiently considered the aforementioned difficulties [88, 96, 95]. With the latter two references [96, 95], the wavelength dependence of the thermo-optical parameter can be estimated, which further supports the choice of a very low, but positive value of $\partial n/\partial T$ (Fig. 7.15). Since reliable values for the photo-elastic parameters are even harder to find than for the thermo-optical parameter, and because the numerical model yields a higher error for the induced stresses in the gain medium than for the temperature distribution, the thermo-optic and photo-elastic effect were combined in one parameter. This was done by considering the thermo-optic parameter $\partial n/\partial T$ as a fit parameter with values based on $\partial n/\partial T = +0.4 \cdot 10^{-6} \mathrm{K}^{-1}$ (see table 7.3). Further parameters which vary significantly in the literature are the thermal expansion coefficients α. Many authors specify these coefficients as a tensor α_{ij}'' in the principal coordinates for thermal expansion X_1', X_2', and X_3'. This tensor can then be transferred to other coordinate systems, for instance, into the thermal expansion coefficients in the refractive index system α_{mpg}. For example Pujol et al. [86] and Biswal et al. [98] specify the thermal expansion coefficients as follows in the principal system of thermal expansion, and in the principal system of the indicatrix, respectively:

$$\alpha_{ij}'' = \begin{pmatrix} 10.64 & 0 & 0 \\ 0 & 2.83 & 0 \\ 0 & 0 & 23.44 \end{pmatrix} \cdot 10^{-6} \ K^{-1}, \ \alpha_{mpg} = \begin{pmatrix} 11 & 0 & 7.1 \\ 0 & 2.4 & 0 \\ 7.1 & 0 & 17 \end{pmatrix} \cdot 10^{-6} \ K^{-1}.$$

In the following we used the approximate values for the thermal-expansion coefficients $\alpha_a = 4 \cdot 10^{-6} \mathrm{K}^{-1}$, $\alpha_b = 3.6 \cdot 10^{-6} \mathrm{K}^{-1}$, and $\alpha_c = 8.5 \cdot 10^{-6} \mathrm{K}^{-1}$ along each crystallographic axis as given by the crystal manufacturer and by Mochalov [79, 80]. Furthermore, we assumed the a-axis to be approximately parallel to the N_m-axis, and the c-axis parallel to the N_g-axis.

Due to the huge uncertainties for many material parameters, as well as the necessity to consider some properties as fit parameters, the simulation outputs can only serve as a rough prediction. Nonetheless, they can help to support explanations based on experimental evidence. All material parameters of Yb:KGW and all fit parameters that were used in the following simulations are listed in table 7.3. Furthermore, the temperature dependence of the thermal conductivity was taken into account and modeled with a 1/T behavior as expected for temperatures above 100 K [99], and as supported by the results obtained by Aggarwal et al. [100].

In order to validate the numerical model and the employed material parameters (table 7.3), we calculated the thermal lens induced in the Yb:KGW crystal by a fiber-coupled 6.7 W laser diode with a beam quality of $M^2 \approx 24 \times 24$. The unpolarized light of the diode was focused to a circular pump spot of a diameter of 110 μm. The maximum absorption amounted to 75% due to the different polarization of the pump light. Fig. 7.16 shows a comparison between the numerical results and the

7.3. Femtosecond 5 W Yb:KGW slab oscillator

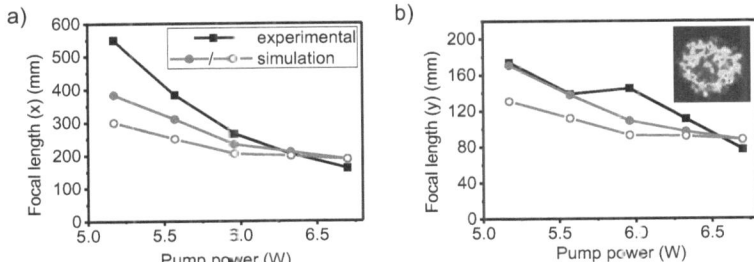

Figure 7.16.: Focal length of the thermal lens induced in the Yb:KGW crystal by pumping with a unpolarized fiber-coupled 6.7 W diode. The black squares show the experimentally obtained values, whereas the red circles show the numerical results (hollow symbols: varying laser efficiency, full symbols: fixed efficiency)

experimentally obtained values. The inset shows an image of the beam profile of the fiber-coupled diode taken with a CCD-camera. The experimental values were obtained by measuring the divergence angle of the output beam and comparing the results to ray-matrix calculations as described in section 7.3.4. The simulations were done for two different scenarios. In the first calculation (hollow symbols), the actual measured laser efficiency was taken into account, i.e. the heat load was considered as the amount of pump power that was not converted to laser light minus additional 5% accounting for further losses, such as reflections on the crystal surfaces and lenses, non-ideal mirrors and so on. The other calculation was done for a fixed rate of pump light converted to heat. Here, the laser efficiency at maximum laser power plus the additional 5% were taken into account. In both calculations, the beam profile of the pump beam was considered as a Gaussian distribution with the M^2-factor of this diode taken into account. Considering the huge uncertainties in the material parameters of Yb:KGW, the simulations predict the behavior of thermal lens sufficiently well. The simulations with a fixed rate of pump light converted to heat seem to fit better to the experimental values. This behavior was also noticeable with simulations of broad-area diodes, as shown later. This behavior could be due to a stimulated excitation of anti-Stokes Raman scattering which leads to a conversion of pump light to fluorescence light instead of heat [101]. Yb:KGW is known to show strong Raman signals [88]. Another reason for the deviation for low pump powers (here below ~5.5 W) could be that some part of the laser losses are due to

Table 7.4.: Laser parameters for each level of pump power that were assumed for the simulation of the thermal lens induced by a 12 W broad-area emitter with $M^2 = 50 \times 1.2$. All values were determined experimentally during cw-operation of the laser at a constant pump-beam diameter of $D = 400 \times 200$ μm^2 for all cases. The rate of pump light that is converted to heat, $\eta_{P \rightarrow heat}$, was considered as the difference between pump and laser power minus an estimated additional loss of 5%.

P_{pump} (W)	P_{laser} (W)	λ_{pump} (nm)	$\eta_{P \rightarrow heat}$	Abs. coeff. (m^{-1})
12.0	5.28	981.0	0.51	380
11.4	4.54	980.4	0.55	344
10.7	3.66	979.4	0.61	295
10.0	2.58	978.8	0.69	254
9.2	1.51	978.0	0.79	198

re-absorption of laser light. This absorption would be homogeneously distributed over the full crystal length as opposed to the exponentially decaying pump light. However, since the simulation model remains relatively simple and many material parameters are not considered or imprecisely known, this numerical model should not be stressed too much and only be considered as a guideline for the laser design.

As discussed in section 7.3.4, one can notice effects that suggest a strong influence of the specific beam profile of a broad-area diode on the thermal effects induced in the gain medium. In order to further investigate this observation, the thermal effects in the Yb:KGW crystal induced by a 12 W broad-area diode with a beam-quality of $M^2 \approx 50 \times 1.2$ were simulated. Fig. 7.17 shows the results of the simulations (red circles) in comparison to the experimental results (black squares). The simulations were performed for three different settings. Each setting was additionally calculated for a fixed and an adapted conversion rate of pump power to heat. The input parameters for each pump power level were chosen according to experimentally observed values. The used parameters are summarized in table 7.4. If the conversion rate of pump power to heat was assumed to be fixed, a value of $\eta_{P \rightarrow heat} = 0.51$ was chosen. Fig. 7.17 a) shows the numerical results that were obtained by assuming a Gaussian intensity distribution of the pump light with the beam diameters of $D = 400 \times 200$ μm^2 as calculated by the ray-transfer matrix method and $M^2 = 50 \times 1.2$. This is the pump light distribution as predicted by calculations with the diode specifications. Besides that, experiments showed that

7.3. Femtosecond 5 W Yb:KGW slab oscillator

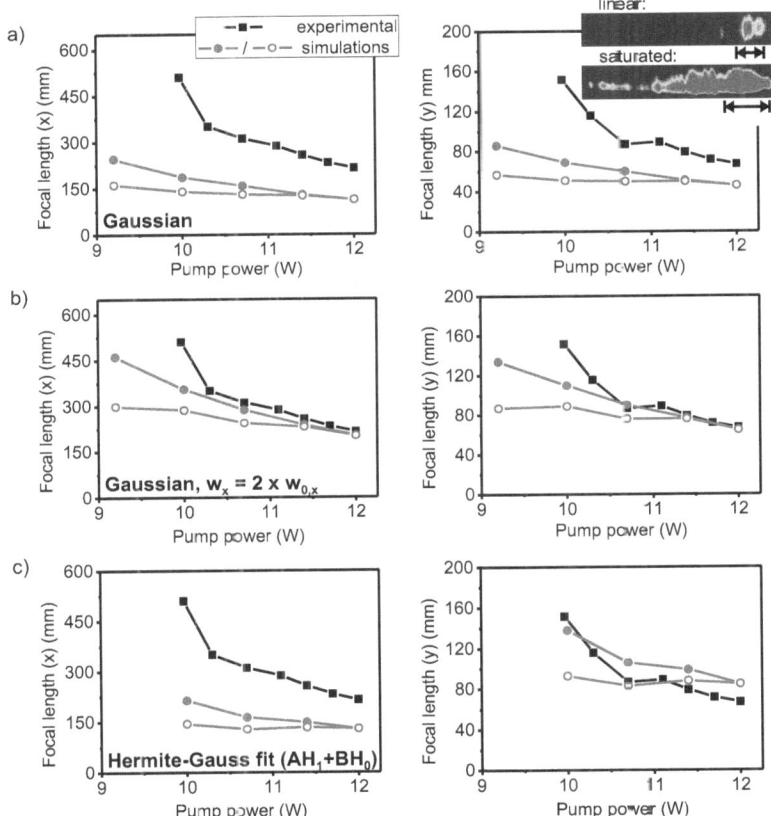

Figure 7.17.: Comparison of experimentally obtained values of $f_{thermal}$ (black squares) with numerical results (red circles). The thermal lens was induced by a 12 W broad-area diode. The full red circles depict simulations with a fixed rate of pump light converted to heat, whereas for each hollow symbol, the measured optical-to-optical efficiency was taken into account individually. The experimental values were compared to three different numerical models: a) Model with a Gaussian pump light distribution with beam diameters calculated according to the diode specifications. b) A heuristic approach with a Gaussian pump light distribution as in a), but twice the horizontal beam diameter. c) A Model with the pump beam profile described by a superposition of two different Hermite-Gaussian beams.

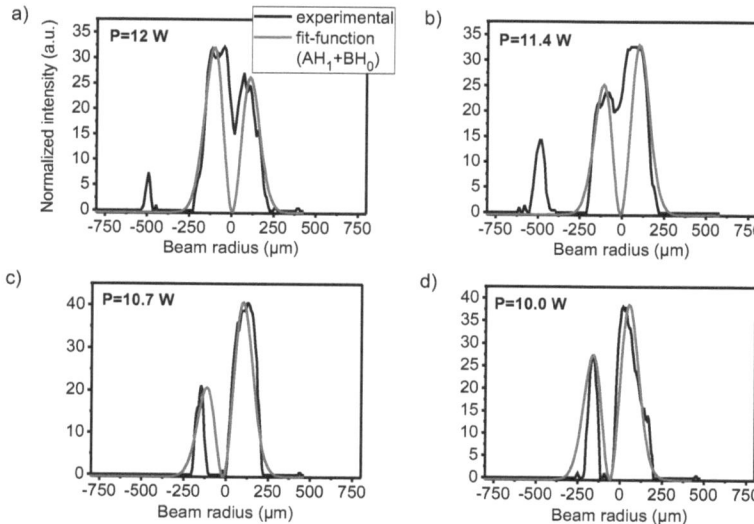

Figure 7.18.: Comparison of experimentally measured beam profiles of a 12 W broad-area diode (black lines) with fitted superpositions of two Hermite-Gaussian modes (red lines).

the diameters of the Gaussian distribution obtained in this way, are also relevant for calculating the necessary laser beam diameters. For example, for the most efficient cw-operation with a diffraction limited laser beam, the laser mode should be of approximately the same size as the calculated diameter along the slow axis and slightly smaller than the calculated diameter along the fast axis. Those are general guidelines which are valid for pump diodes with a nearly diffraction limited beam profile along the fast axis and an approximately flat-top beam profile along the slow axis. Horizontally increasing the pump mode in the experiment to twice the horizontal laser mode diameter, resulted in a value of $M^2 \approx 2$ for the laser beam along the horizontal direction. However, the focal lengths that are predicted by the simulation shown in Fig. 7.17 a) are significantly shorter than the experimental values. Thus, two different approaches were examined. A merely heuristic approach is shown in Fig. 7.17 b). It shows the same comparison, however, with an assumed Gaussian pump-light distribution with twice the horizontal diameter, i.e. a diameter of $D = 2 \cdot 400 \times 200 \ \mu m^2$ and $M^2 = 50 \times 1.2$. This would be misleading for the de-

7.3. Femtosecond 5 W Yb:KGW slab oscillator

Table 7.5.: Fit parameters A, B, and $w_{x,0}$ for each power level that were used to model the horizontal beam profile by a superposition of zeroth and first order Hermite-Gaussian modes.

P_{pump} (W)	$w_{x,0}$ (μm)	A	B
12.0	150	2.0	-0.2
11.4	150	1.5	+0.2
10.7	150	0.6	+0.2
10.0	150	1.2	+0.2

sign of the laser resonator regarding optical properties. Nevertheless, the numerical results fit the experimental results quite well for pump powers above ~ 10 W. The deviations for lower pump powers could be due to the same reasons as discussed above for pumping with the fiber-coupled diode. This heuristic approach of twice the horizontal beam diameter was suggested by considering the beam profile of a broad-area diode. The insets in the upper left corner of Fig. 7.17 show CCD-camera images of the beam profile for a pump power of 12 W. The upper image shows a linear plot with the diameter of the assumed Gaussian beam indicated by the small black arrow. A saturated image reveals the full details of the beam profile. Probably the thermal effects caused by the pump-beam cannot be correctly described by assuming a perfect Gaussian distribution, whereas the influence of the pump beam profile on the laser mode is still be predicted correctly with the results of ray-matrix calculations with the M^2-factor taken into account. In order to develop a more physical assumption for the simulations, the pump beam intensity profiles were modeled in the horizontal direction by a superposition of Hermite-Gaussian modes of the first and zeroth order [48, 102]:

$$I(x,y,z) = C \cdot \left(A \cdot H_1\left(\frac{\sqrt{2}x}{w_x(z)}\right) + B \cdot H_0 \right)^2 \cdot \exp\left(-\left(\frac{2x^2}{w_x(z)^2} + \frac{2y^2}{w_y(z)^2}\right)\right), \quad (7.2)$$

whereas H_0 is the fundamental order Hermite-polynomial $H_0 = 1$, H_1 is the first order polynomial $H_1(x) = 2x$, A and B are fit parameters, and C is a normalization constant to ensure that $\int\int I(x,y,z)dxdy = P$:

$$C = \frac{P}{\pi w_x(z) w_y(z)} \cdot \frac{1}{A^2 + B^2/2}. \quad (7.3)$$

Besides the parameters A and B, also the minimum pump beam waist $w_{x,0}$ inside the laser crystal was considered as fit parameter. The combinations used of all three parameters for each power level are summarized in table 7.5. The diagrams in Fig. 7.18 show the modeled horizontal beam profiles (red curves) in comparison to the experimental profiles (black curves) for four different levels of pump power. The results of the numerical simulations of the thermal lens with these fitted beam profiles are shown in Fig. 7.17 c). The agreement with the experimental results is better compared to the simulations with a regular Gaussian intensity distribution (Fig. 7.17 a). However, compared to the simulations with twice the horizontal beam diameter (Fig. 7.17 b), the results still show larger deviations. These deviations might be partially due to the higher numerical errors compared to simulations with Gaussian intensity distributions because of the irregular beam profile. Nonetheless, this model of a superposition of two Hermite-Gaussian modes for the horizontal beam profile is most likely still too simple and can be improved by taking further higher order modes into account.

In conclusion, the outcome of these simulations can be taken as a strong indication that for high-power broad-area diodes the distinct peaks of their beam profile result in a significantly different thermal lens compared to diodes with Gaussian or top-hat beam profiles. Thus, these peaks in the beam profile need to be taken into account for valid simulations of the thermal lens induced by broad-area diodes. Opposed to fiber coupled diodes for example, assuming a Gaussian intensity distribution for simulations of the thermal effects induced by broad-area diodes, resulted in a bad agreement of numerical and experimental results of the focal lengths of the thermal lens. Further work still needs to be done in order to improve this numerical model of the thermal lens. Nonetheless, the approach of modeling the horizontal beam profile as a superposition of higher order modes is promising to lead to a method which can derive the thermal lens induced by broad-area diodes by measuring their specific beam profile.

7.3.6. Comparison between N_g-cut and athermal orientation of Yb:KGW for the slab-laser geometry

A very nice feature of Yb:KGW is the possibility of the so-called athermal orientation of the laser crystal which can reduce the thermal lens tremendously [98]. For such a crystal orientation, a negative value of $\partial n/\partial T$ can exactly compensate the influence of the thermal expansion and bulging of the laser crystal. Therefore, the thermal lens can be very weak over a wide range of pump powers. For the athermal orientation,

7.3. Femtosecond 5 W Yb:KGW slab oscillator

the following condition needs to be fulfilled [98]:

$$\partial n/\partial T + (n-1)\cdot \alpha = 0, \tag{7.4}$$

with the index of refraction n, the thermo-optic parameter $\partial n/\partial T$, and the thermal-expansion coefficient α along the direction of polarization, respectively. One can notice that for a laser light polarization under a specific angle with respect to the N_g-axis, the thermo-optic parameter can become negative while the thermal-expansion coefficients remain positive (Fig.7.15). A realization of an Yb:KGW laser with the laser crystal in athermal orientation was successfully shown by Hellström et al. [103]. The N_g-cut crystal orientation, as used in this work, does not allow for an athermal orientation because even though the thermo-optic parameter is very low, it is still positive. However, with the N_g-cut orientation it is possible to remove the heat mainly along the b-axis which shows the highest thermal conductivity while simultaneously orienting the laser polarization closer parallel to the a-axis than it would be possible for the athermal orientation. Since the emission cross sections are highest along the a-axis, the N_g-cut orientation allows for the lowest heat load due to efficient heat extraction through heat conduction on one hand, and most efficient laser operation on the other hand. In conclusion, both crystal orientations are well suited to build efficient longitudinally diode-pumped high-power bulk lasers. The N_g-cut orientation allows for a more efficient heat removal from the crystal while the athermal orientation takes a stronger heating into account but can compensate the thermal effects.

In this work, we decided for the N_g-cut orientation in order to find a crystal geometry which allows for a practical experimental setup but yet meets all of our requirements such as support of a flat Brewster-cut geometry and efficient heat removal. In addition to that, to our experience, it is more important to most efficiently remove the heat from the laser crystal and to try keeping the thermal lens as close to an ideal lens as possible by employing a thoroughly designed crystal cooling. Then the thermal lens can be taken into account in the laser resonator design. Even if the thermo-optic effect and the thermal expansion can compensate each other with an athermally oriented crystal, still higher order optical aberrations can appear, for example due to the photo-elastic effect. By employing the N_g-cut orientation, we could achieve a nearly diffraction limited laser mode for pump powers of up to 18 W (M^2<1.1), while in comparable athermal setups, for example ref. [103], the beam quality was limited to $M^2 \approx 1.5$.

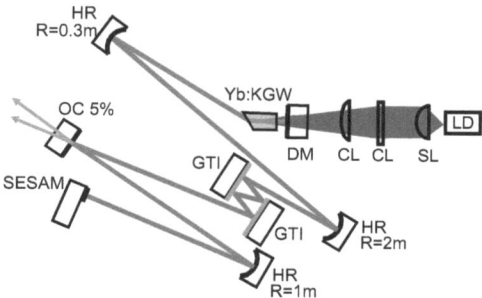

Figure 7.19.: Schematic of the laser setup: LD denotes the pump diode; SL and CL represent an aspherical and two cylindrical lenses, respectively. HR are highly reflective mirrors, DM is a dichroic mirror, GTI are dispersive mirrors, SESAM is a saturable absorber mirror, and OC denotes the output coupling mirror.

7.3.7. Experimental setup of the femtosecond 44 MHz oscillator

The experimental setup of the mode-locked 44 MHz oscillator is schematically shown in Fig. 7.19. As pump diode, a 200 μm broad-area diode capable of providing 17.9 W at 981 nm was employed. The pump optics were identical to the setup described in section 7.3.4. The pump light was first collimated by an anti-reflection coated aspherical lens with a focal length of $f = 3.1$ mm and refocused into the Yb:KGW crystal by two spherical cylindrical lenses with a focal length of $f = 200$ mm and $f = 50$ mm, respectively. For this set of lenses and the particular 18 W broad-area diode, the pump light was focused to a diameter of approximately 370×190 μm² inside the laser crystal. This estimation is based on ray-transfer matrix calculations and agrees well with the experimental values for the optimum laser mode diameter. The losses from all three lenses summed up to approximately 9% of the pump light. The flat side of the laser crystal was anti-reflection coated for both the laser- as well as for the pump-wavelength around 1027 nm and 981 nm, respectively. The Brewster-cut crystal surface was uncoated. The laser mode was focused into the laser crystal with a curved mirror with a radius of curvature of $R = 300$ mm, resulting in a beam diameter of approximately 400×200 μm². The laser resonator was extended by a second curved mirror with a radius of $R = 2000$ mm to achieve an overall resonator length of 3.4 m, corresponding to a repetition rate of 44 MHz. Finally the laser beam was focused onto the SESAM by another curved mirror with

7.3. Femtosecond 5 W Yb:KGW slab oscillator

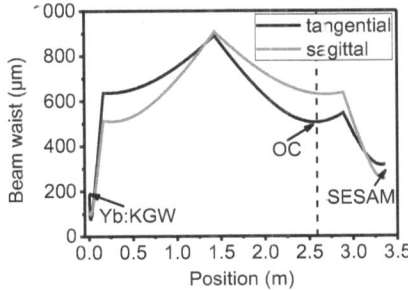

Figure 7.20.: Linear ray-transfer matrix calculations of the beam waist at each position inside the laser resonator. Yb:KGW: position of the gain medium which also acts as one resonator end-mirror, SESAM: position of the saturable absorber mirror which acts as the second end-mirror, OC: approximate position of the output coupling mirror.

$R = 1000$ mm.

The SESAM acted as the second resonator end-mirror. An additional mirror was used output coupling mirror. For all experiments in the mode-locked regime, we used an output coupling mirror with a reflectivity of $R = 95\%$, resulting in an effective output coupling loss of 10% per round-trip (Fig. 7.19). A single output beam could be obtained by coupling out through the dichroic mirror at the pump side of the laser crystal. Another possibility to achieve only one output beam would be to employ a V-shaped resonator design where the beam path inside the crystal is folded in a V-shape. Such a V-shaped resonator design could be similar to reference [91]. However, the solution with a V-shaped resonator increases the accumulation of self-phase modulation per round-trip. Therefore, employing a dichroic coating seems to be the most straightforward method. The thermal lens was taken into account and the resonator length was adjusted in order to achieve a repetition rate of 44 MHz. The beam waist along each position in the resonator for cw-operation is shown in Fig. 7.20. The positions of the gain medium and the saturable absorber mirror are marked with Yb:KGW and SESAM, respectively. Furthermore, the approximate position at which the output coupling mirror was inserted is indicated. In mode-locked operation, nonlinearities in the gain medium influence the resonator properties. The primary nonlinearity is the optical Kerr effect which leads to the formation of a Kerr lens. Hence, the laser resonator needs to be adjusted to compensate for this lens-like effect. The compensation of a Kerr-lens for the chosen resonator can easily be

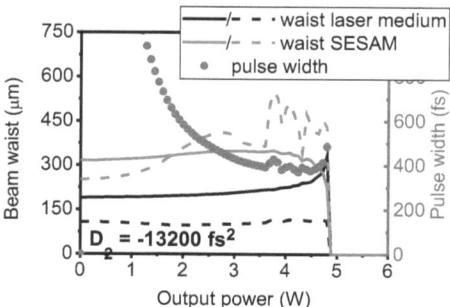

Figure 7.21.: Simulation of the output power dependence of the beam waist inside the gain medium (black lines), as well as on the SESAM (gray lines). The beam waists are shown for the tangential (solid lines) and the sagittal plane (dashed lines), respectively. The expected pulse width is depicted by the red squares. The calculation were performed for an inserted GDD of -13200 fs^2 per round-trip. The deviations to the experimental results are due to numerical problems.

achieved by reducing the distance of the SESAM to the curved mirror. For a suitable resonator it is beneficial if it is insensitive with respect to lens effects inside the gain medium. In particular when employing broad-area pump diodes, such a resonator design allows for a simpler replacement of the pump diode if necessary. Fig. 7.21 and Fig.+7.22 show simulations of the beams waists inside the gain medium (black lines), as well as on the SESAM (gray lines) in dependence on the output power. The simulations were done for two different amounts of inserted GDD and for both planes, tangential (solid lines) and sagittal (dashed lines), respectively. Furthermore, the distance between SESAM and the curved mirror was decreased by 10 mm over the output power range of 0 W to 7 W. The expected pulse width is indicated by red squares. The simulations were done by the extended ray-transfer matrix method as described in Chapter 5.2. Due to numerical problems in our model as discussed in Chapter 5.2, also the huge nonlinear index of Yb:KGW of $n_2 = 20 \cdot 10^{-16} \text{cm}^2/\text{W}$ leads to numerical instabilities and discrepancies compared to the experimental results. However, these simulations could still be used to find a suitable resonator for mode-locking. The exact laser parameters were determined experimentally.

The used laser resonator was designed to operate in the middle of the stability regime, i.e. Kerr-lens mode-locking did not play a dominant role for pulse generation. Therefore, in order to operate the laser in the solitary mode-locking

7.3. Femtosecond 5 W Yb:KGW slab oscillator

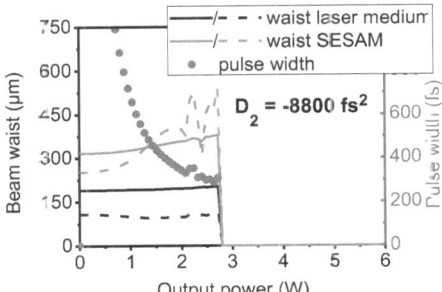

Figure 7.22.: Simulated output power dependence of the beam waist inside the gain medium (black lines), as well as on the SESAM (gray lines) for an inserted GDD of -8800 fs² per round-trip. The expected pulse width is depicted by the red squares. The resonator instability starting around ~3 W output power are due to numerical problems.

regime [34, 35], net negative dispersion (GDD) per round-trip was introduced into the resonator by Gires-Tournois interferometer (GTI) mirrors, and a saturable absorber mirror [36, 37, 39] was introduced as one end-mirror. The mode size on the SESAM was estimated to be approximately 600×800 µm². Two different SESAMs were used. The first SESAM had a saturable absorption (modulation depth) of $\Delta R = 0.4\%$ and a saturation fluence of 120 µJ/cm². The second SESAM had a slightly higher saturable absorption of $\Delta R = 0.5\%$ and a saturation fluence of 70 µJ/cm². The employed saturable absorbers are commercially available low-finesse anti-resonant SESAMs designed for a center wavelength of 1030 nm and 1070 nm, respectively. All SESAM parameters in this work are given according to the manufacturer's specifications. Both SESAMs were able to stably mode-lock the laser. However, the second SESAM was operated at the limit of its design wavelength-range which lead to laser pulses which were longer than the transform limit, as will be discussed later. The pulse width could be modified by changing the number of bounces on the GTI mirrors and thus varying the amount of negative dispersion per round-trip.

7.3.8. Mode-locked properties of the Yb:KGW oscillator

The maximum output power in pulsed operation did depend on the pulse width and therefore on the amount of GDD per round-trip as well as on the saturable

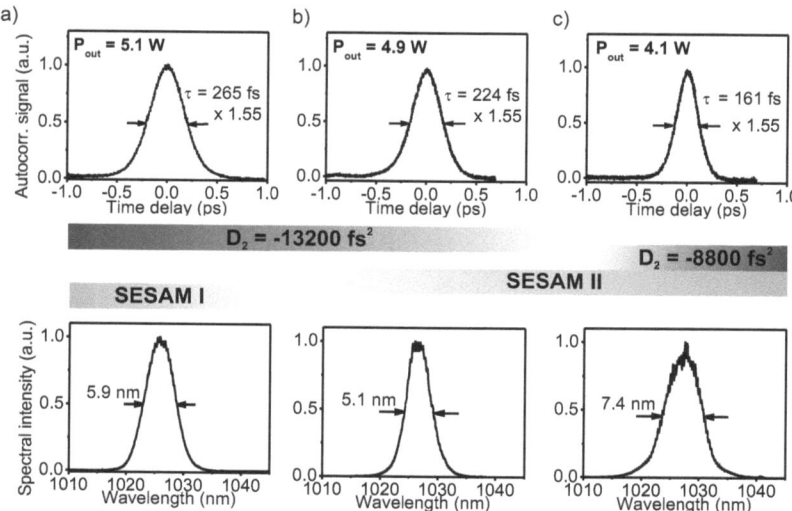

Figure 7.23.: Autocorrelation traces and spectra measured for three different laser configurations. Figs. a) and b) show measurements for the same amount of inserted GDD but two different SESAMs. Figs. b) and c) show the results for the same SESAM but a different amount of GDD. The SESAM parameters were $\Delta R = 0.5\%$, $F_{sat} = 70\mu J/cm^2$, $\lambda_{design} = 1070$ nm (SESAM I) and $\Delta R = 0.4\%$, $F_{sat} = 120\mu J/cm^2$, $\lambda_{design} = 1030$ nm (SESAM II), respectively. The differences between these three configurations are discussed in the text.

losses (modulation depth) and GDD of the employed SESAM. Fig. 7.23 shows three examples for different amounts of dispersion and different modulation depths of the SESAM. The maximum output power of 5.1 W at a pulse width of 265 fs and a repetition rate of 44 MHz was achieved at an inserted GDD of approximately -13200 fs² and a SESAM with a modulation depth of $\Delta R = 0.5\%$ (Fig. 7.23 a). The pump power was 17.8 W, resulting at an optical-to-optical efficiency of 29%. With the same amount of GDD but a different absorber with $\Delta R = 0.4\%$ modulation depth, the pulse width decreased to 224 fs at an output power of 4.9W (Fig. 7.23 b). Using the same SESAM but changing the GDD to -8800 fs², did result in the shortest measured pulse width of 161 fs at an output power of 4.1 W. This power was obtained for a pump power of 17.6 W, which results in an optical-to-optical efficiency of still more than 23%. The beam quality in all cases was measured to be

7.3. Femtosecond 5 W Yb:KGW slab oscillator

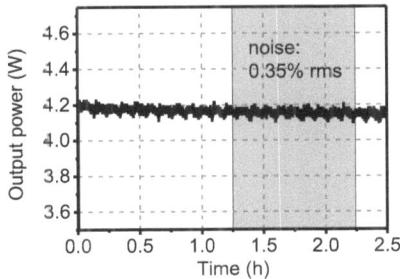

Figure 7.24.: Measurement of the output power for a laser configuration with an inserted GDD of -13200 fs^2 per round-trip and the SESAM with a modulation depth of $\Delta R = 0.5\%$. The intensity noise was determined to be 0.35% rms over one hour.

diffraction limited with a M^2 value of less than 1.1.

The laser did start in mode-locked but in cw-operation. However, mode-locking could be easily initiated by a gentle tap on the optical table, and after initiation it remained stable for hours. The output power fluctuations were measured to be 0.35% rms over one hour (Fig. 7.24).

In all setups in which the SESAM with $\Delta R = 0.4\%$ was used, the maximum output power was limited by the appearance of cw-breakthroughs. In the case with the SESAM with a slightly higher modulation depth of $\Delta R = 0.5\%$, the output power was only limited by the available pump power. However, while the pulses with the 0.4% saturable absorber were transform limited with a time-bandwidth product of 0.33-0.34, the setup with the 0.5% absorber showed a slightly increased time-bandwidth product of 0.45 (Fig. 7.23). This increase was due to the operation of this SESAM very close to the limit of its design wavelength range. According to the specifications of the manufacturer, the lower wavelength limit of this SESAM is around 1025 nm. Due to the also higher non-saturable losses for this wavelength, the laser wavelength was additionally shifted from 1027 nm towards 1025 nm.

Fig. 7.25 shows a typical dependence of the output power and the efficiency on the pump power for the case with a SESAM with a low modulation depth of 0.4%. The introduced GDD in this case was -13200 fs^2 per round-trip. The distance between SESAM and the curved mirror was changed while increasing the pump power. Close to the laser threshold around 15.7 W the laser operated in the cw-regime. For

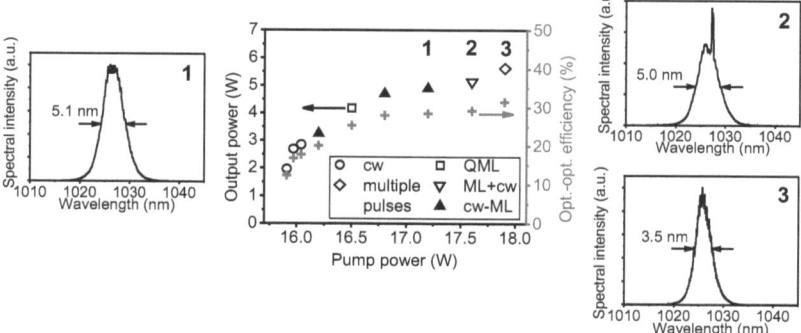

Figure 7.25.: Typical measurement of the output power (circles, triangles, squares, and diamonds) and optical-to-optical efficiency (crosses) for the case of a low modulation depth of $\Delta R = 0.4\%$. The amount of inserted GDD accounted to -13200 fs^2 per round-trip. The insets show the laser spectrum for various pump powers and operation regimes: 1: spectrum for cw-mode-locking, 2: cw-breakthroughs for higher pump powers, 3: narrowing and modulation of the spectra for high pump powers. The region of pure cw-mode-locking can be expanded to the maximum pump power by slightly increasing the SESAM modulation depth.

increasing pump powers, we observed a small region of Q-switching followed by a broader region of stable cw-mode-locking. Further increasing the pump power above ~17.2 W lead to cw-breakthroughs or to a narrowing and modulation of the laser spectrum. This narrowing and modulation was due to a break-up into multiple pulses. As mentioned above, the upper pump power limit for cw-breakthroughs could successfully be extended to the full pump power by employing a SESAM with slightly higher modulation depth. This extended the range of pure cw-mode-locking up to the full pump power.

All measured laser spectra showed a slight modulation which could be attributed to an etalon-like effect due to spurious reflections from the parallel surfaces of the laser crystal and the dichroic mirror with an approximately 1.2 mm wide gap between them. This effect also reduced the efficiency because, in order to mode-lock the laser, the laser crystal needed to be slightly rotated off the optimum orientation in respect to the main refractive axes. In this way the Fabry-Perot or etalon like effect can be reduced which facilitates mode-locking [104]. Since the laser crystal was cut without a wedge, a small angle did arise between the propagation axis and the optical axis which thus leads to slightly increased losses due to polarization rotation. Therefore, the laser efficiency could be further increased by taking this effect into account, for instance, by cutting the laser crystal with a small wedge. However, the B-integral (Eq. 4.8) for $P_{out} = 5$ W output power and a pulse width of $\tau = 224$ fs accounted to $B = 4.87$. Therefore, for a further increase of the output power, for example by employing a second pump diode, maybe a stability limit can be reached. If this will be the case, perturbations due to nonlinear effects can be avoided by increasing the mode diameter inside the gain medium or by increasing the repetition rate.

The laser was successfully used for second harmonic generation. In a 4 mm long LBO crystal ($\theta = 90°$, $\phi = 13.8°$), a conversion efficiency of more than 50% was achieved. One of the two output beams was sent through a Faraday isolator, a half-wave plate to adjust the polarization, and focused into the LBO crystal by an achromatic lens with a focal length of $f = 50$ mm. Approximately 2.1 W incoming laser power onto the LBO crystal was converted to 1.1 W at a wavelength of $\lambda_{SHG} = 512$ nm.

7.4. Generation of high-power femtosecond supercontinua

7.4.1. Supercontinua generated by tapered fibers

The laser was successfully used to generate supercontinua with tapered fibers. One of the two output beams was sent through a half-wave plate, a Faraday-isolator, and then coupled into a tapered fiber. The laser generated pulses of approximately 224 fs width. The output power in one beam was 2.4 W. After the Faraday-isolator up to 2.1 W remained. The laser beam was coupled into the tapered fiber by a $10 \times$ microscope objective with an NA of 0.3. Fig. 7.26 shows the spectral output for three tapered fibers with different taper waist diameters between 3-5 μm. The waist lengths of all fibers was approximately 6 cm. With 2.1 W pump power up to 1.3 W average whitelight power could be achieved with 5.0 μm waist diameter. The spectrum ranged from approximately 650 nm to 1450 nm, resulting in a spectral intensity of more than 1.5 mW/nm. By decreasing the waist diameter to 3.8 μm the spectrum becomes broader. However, the coupling losses in the taper region increase so that the overall average power decreases to 1.1 W. Very broad spectra

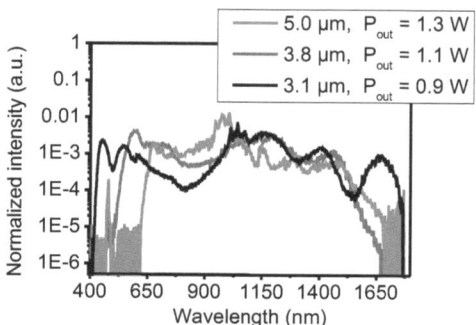

Figure 7.26.: Supercontinua generated with 224 fs pulses. An input power between 1.9 W and 2.1 W was coupled into different tapered fibers with 3.1 μm (solid line), 3.8 μm (dashed line), and 5.0 μm (dash-dotted line) waist diameter and 90 mm waist length. The overall average whitelight output power was 0.9 W, 1.1 W, and 1.3 W.

between 420 nm to over 1750 nm with an overall average power of 0.9 W at 1.9 W input power were achieved by decreasing the waist diameter to 3.1 μm (see black curve in Fig. 7.26).

As a further experiment, spectrally filtered sections of supercontinuua were investigated. In order to do this, different interference filters were placed behind the fiber output. The stability of a spectrally narrow filtered section of a supercontinuum is dependent on the filter central wavelength and filter bandwidth. Fig. 7.27 shows a measurement of the intensity noise of 10 nm broad filtered sections of a supercontinuum generated by coupling approximately 1 W average power at 265 fs into a tapered fiber with 2.9 μm waist diameter. At an overall average power of 500 mW this specific combination is suitable to generate high spectral intensities of several mW/nm around 500 nm and in the near infrared. In order to characterize the intensity noise, the power of the spectrally filtered light was measured over 10 minutes and the noise was derived in relation to the average power of the filtered light. The spectrum was first filtered with either a short-pass or a long-pass filter. Then the specific wavelength was selected with a band-pass filter. The intensity measurements were performed with a photo-detector with a 3dB-bandwidht of $f_{3dB} = 1$ MHz (New Focus, model 2031). In the near infrared for wavelengths longer than the laser wavelength (around 1100-1500 nm), the noise was measured to range between 0.5% to 2% rms. The intensity fluctuations increased to 3-4% rms

7.4. Generation of high-power femtosecond supercontinua

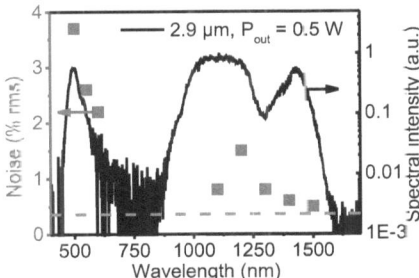

Figure 7.27.: Measurements of the intensity noise of 10 nm broad spectrally filtered sections of a supercontinuum with high spectral intensities around 500 nm as well as in the near infrared (black curve). The squares denote the noise at different center wavelengths over 10 minutes. The dashed line indicates the noise level of the laser oscillator over one hour.

in the visible regime around 500 nm. A similar behavior of increased noise in the regime of dispersive waves but lower noise around the pump wavelength and in the region of Raman solitons in the infrared was observed before, for instance by Corwin et al. [105]. The reason for this behavior can be attributed to the effects causing the conversion of light at certain wavelengths. While spectral components in the infrared mostly originate from soliton-self-frequency-shifts of fundamental solitons caused by Raman scattering, light at visible wavelengths is generated at the beginning of the SC generation by the emission of non-solitonic radiation during the fission of higher-order solitons [3, 105]. The red-shifted fundamental solitons produce little noise due to the intrinsic stability of their temporal envelope. However, the higher order solitons at the beginning of the SC generation process are noisy, for example because of laser fluctuations or unstable fiber coupling. This noise is then further amplified due to modulation instabilities [105, 106].

7.4.2. Supercontinuum generation in polarization-maintaining PCF

To investigate the possibility of the generation of polarized femtosecond supercontinua, one of the two output beams was coupled into ∼20 cm of polarization-maintaining PCF (Thorlabs, NL-PM-750). In contrast to conventional polarization maintaining fibers with a solid core an induced birefringence by stress rods (PANDA),

98   Chapter 7: Multi-Watt 44 MHz femtosecond supercontinuum source

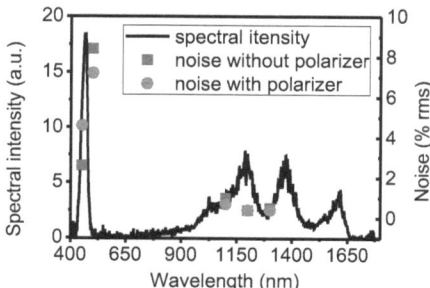

Figure 7.28.: Measurement of the spectral intensity noise of a supercontinuum generated in a polarization-maintaining PCF. The solid black curve shows the achieved spectrum with an approximate output power of 75 mW. The red squares and green circles show results of measurements of the intensity noise of 10 nm spectrally filtered sections of this supercontinuum with an polarizer inserted or removed in front of the photo-detector, respectively.

the birefringence in PCF can be obtained by an asymmetric pattern of holes, see for example ref. [107]. The fiber used in this experiment had a zero-dispersion wavelength of 750 nm, and a core field diameter of $D = 1.8$ μm. Due to the small mode field diameter, the input power had to be attenuated to approximately $P_{in} = 300$ mW. For higher input power, the front facet of the PCF was damaged. After cleaving a damaged facet, the maximum coupling efficiency could be re-produced. By coupling into the fiber with a 60× microscope objective, a conversion efficiency of 25% and a maximum output power of $P_{out} = 75$ mW were obtained. The input light was linear polarized at a center wavelength of 1027 nm and a pulse width of approximately 250 fs. The input polarization was rotated by a half-wave plate parallel to one refractive fiber axis in such a way, that the output beam was also linear polarized. This was tested with a Glan-Thomson polarizer for the case of cw operation and for the laser wavelength. The spectrum of the generated supercontinuum is shown by the black curve in Fig. 7.28. In order to measure the polarization stability of spectrally filtered sections of the output beam, the intensity noise was measured, according to the method described in section 7.4. The intensity noise with respect to the rms power is shown for different wavelengths for a measurement without polarizer (red squares), and behind a polarizer (green circles). The general wavelength dependence of the noise is similar to the results found for tapered fibers (see section 7.4) The difference between the measurements with and without polarizer is very small The bigger differences for visible wavelengths might be due to the overall increased noise in this regime.

The results of these measurements show, that femtosecond supercontinua can be generated with a stable linear polarization of the output beam and even for small spectrally filtered sections. By employing a PCF with a larger core-field diameter, it should be possible to increase the power that can be coupled into such a PCF without destruction of the fiber facet. Also fiber pigtails with larger core-field diameters could help to couple more light into the PCF by providing larger core field diameters on the input facet.

7.5. Conclusion

We presented a novel, simple, and highly efficient design for a multi-Watt femtosecond oscillator capable to generate laser pulses shorter than 200 fs. The development of this new source allowed for the realization of a compact femtosecond supercontinuum source with more than 1 W average power. This research contributes to the current work on slab lasers with Ytterbium doped tungstates. We could show that the N_g-cut orientation of Yb:KGW, which is only little regarded up to now, allows to build efficient high-power femtosecond oscillators and can provide better results than using Yb:KGW in the athermal orientation. Furthermore, we could show that the efficiency of diode-pumped femtosecond slab lasers can be significantly increased by employing high-power broad-area diodes. By pumping the laser with a single broad-area diode, an average power of more than 5 W with an optical-to-optical efficiency of more than 28% was achieved. With a pulse width of 161 fs, the optical-to-optical efficiency was still more than 23% which is to our knowledge the highest value for a diode-pumped femtosecond laser with this pulse-width and output power.

Chapter 8
Summary

In summary, we demonstrated two compact, stable, and cost-effective femtosecond supercontinuum sources. We developed two suitable laser oscillators for this purpose and demonstrated the generation of femtosecond supercontinua with over one Watt average power. Furthermore, a comprehensive study of the properties of such supercontinua was presented. Based on former results of supercontinuum generation with diode-pumped lasers, we focused our investigations on Ytterbium lasers as pump sources for tapered fibers and PCF. The results described in this work can be classified into two fields:

1. Research in the field of diode-pumped sub-200 fs Ytterbium laser oscillators as well as a detailed study of the potential of broad-area diodes as pump sources for such lasers.

2. The characterization of the properties of femtosecond supercontina generated in tapered fibers and PCF with pump wavelengths around 1 μm.

The main part of this work focused on the research in the field of diode-pumped sub-200 fs laser oscillators. At first, a 20 MHz Yb:glass oscillator was developed that can replace the widely used Ti:sapphire lasers. The laser is capable to generate pulses as short as 147 fs and output powers of up to 800 mW. By designing the laser resonator with a Herriott-type multi-pass cell, the setup of the oscillator fit onto a footprint of only 20×60 cm^2. In a second approach, the average power was scaled to multiple Watts. While for pulse durations longer than 200 fs, several tens of Watts of average power can be achieved very efficiently with thin-disk lasers, it is still a challenging task to generate multi-Watt average power laser pulses shorter than 200 fs directly from a laser oscillator. Challenges arise for example due to gain narrowing or nonlinear effects. A concept which recently found increasing attention is the so called slab laser concept. Due to novel gain materials such as Ytterbium doped tungstates as well as advances in laser diode technology, significant progress in multi-Watt average power sub-200 fs laser oscillators has been achieved. In this work, we showed that further progress in simplicity and particularly in inefficiency is possible by two measures. On one hand, using a novel kind of pump diodes, so called high-power broad-area diodes, increases the efficiency tremendously. On the other hand, exploiting the N_g-cut orientation of Yb:KGW allows to pump such laser with up to 18 W power and still achieve a diffraction limited laser beam with $M^2 < 1.1$. This crystal orientation yields even better results than the athermal direction of Yb:KGW.

We started our research in the field of high-power femtosecond slab laser oscillators with an investigation of the thermal effects and thermal lensing in N_g-cut oriented Yb:KGW. This gain medium was pumped with up to 18 W from high-power broad-area diodes. The employed broad-area diodes were capable to emit up to 18 W power from a single emitter with waveguide dimensions of approximately 200×3 μm^2. The beam emitted from those diodes is nearly diffraction limited along the fast axis with $M^2 \approx 1.2$, while along the slow axis, still a value of $M^2 \approx 50\text{-}60$ is achieved. However, these diodes show a particular beam profile with distinct intensity peaks. The ratio of the fraction covered by these peaks to the whole beam profile is much larger compared to the same ratio in the case of fiber coupled diodes or diode bars. We found that this phenomenon results in thermal effects that would not be expected for fiber coupled diodes or diode bars. Even though the mode-matching of laser mode and pump beam can be well predicted by assuming a Gaussian pump light distribution in combination with the appropriate M^2-factor, the thermal effects vary significantly from diode to diode and cannot be described by this way. Regarding this, we demonstrated that still diffraction limited laser beams, optical-to-optical efficiencies of nearly 50% in cw operation and almost 30% in mode-locked operation can be achieved with those diodes. Numerical simulations of the thermal effects were performed by a FEA-analysis. However, even though the predicted results agreed well with the experiment for fiber coupled diodes, the employed model still needs to be improved in order to predict the focal length for a specific beam profile of a broad-area diode. A further improvement and more sophisticated model of the beam profiles could improve these simulations. This may allow for the adaption of a laser resonator to a specific broad-area diode from a measurement of a beam profile.

The obtained insights in the thermal effects induced by high-power broad-area diodes were used for the development of a 44 MHz slab-laser oscillator. The oscillator is pumped by a single 18 W broad-area diode and is capable to produce average powers of more than 5 W with an optical-to-optical efficiency of more than 28%. With a decreased pulse duration of 161 fs, the optical-to-optical efficiency is still higher than 23%. This is to our knowledge the highest value for a diode-pumped femtosecond laser with this pulse-width and average power. The efficiency should be further increased by using a laser crystal with a more suitable coating. This helps to suppress spurious reflections and allows to improve the necessary misalignement of the dichroic mirror. Furthermore, a dichroic output coupling mirror can be used to combine both output beams.

By combining tapered fibers and PCF with the developed laser oscillators, two different femtosecond supercontinuum sources were realized. A very compact and flexible 20 MHz source with up to 400 mW average power and a 44 MHz system capable to deliver over 1 W supercontinuum average power. The influence of the taper waist diameter as well as the influence of a pulse pre-chirp on the spectral shape were investigated. Furthermore, we studied the noise properties of supercontinua and a possibility for polarized femtosecond supercontinua. This is to our knowledge the first comprehensive study of supercontinuum properties for a pump wavelength around 1 μm. Theoretical predictions of the influence of the fiber geometry on the spectral shape of the supercontinua were confirmed. Furthermore, the influence of a pulse pre-chirp and of the fiber pig-tails was found to be negligible for supercontinuum generation in tapered fibers. This result differs from the results found for Ti:sapphire wavelengths. We explained this behavior by the greater proximity to the zero-dispersion wavelength of SMF28 fibers as well as the higher anomalous GVD in the waist region for wavelengths around 1 μm compared to 800 nm. For many applications, the intensity noise of spectrally filtered sections of supercontinua is important. In the infrared regime, an intensity noise of less than 2% rms (in most cases even less than 1% rms) was measured, whereas the noise was slightly increased in the dispersive wave regime. This behavior could be attributed to the soliton dynamics during the generation process of supercontinua. The possibility to generate polarized femtosecond supercontinua was investigated by employing 20 cm of polarization-maintaining PCF. In the near infrared region, the polarization fluctuations were found to be negligibly small compared to the overall intensity noise of this section. However, in order to increase the input power, measures need to be taken in order to protect the facets of such PCF from damages due to high intensities. This could be achieved by using fibers with a larger mode-field diameter or by splicing large mode area PCF pig-tails and nonlinear PCF together.

The results of this thesis will support future developments of sub-200 fs high-power laser oscillators. For example, the specific 44 MHz laser oscillator developed in this work can be improved by combining both output beams by employing a dichroic output coupling mirror. In addition, pumping with several pump diodes can increase the output power. Several diodes could be combined by polarization- or wavelength-multiplexing. Furthermore, the results found are also useful to improve other laser designs. The N_g-cut crystal orientation in combination with broad-area diodes can be applied to other resonator geometries, e.g. with the gain medium in the center of the resonator. Besides the significance for the design of laser oscillators, the possibility to generate femtosecond supercontinua with over 1 W average power creates the opportunity for many future applications. For instance, an optical parametric amplifier (OPA) pumped with nanojoule laser pulses could be realized. This idea would benefit from the significantly higher spectral seed intensities that are possible with supercontinua generated in tapered fibers compared to those generated in sapphire plates. When supercontinua are generated with more than 1 W average power, still 2.5 W laser power are left that can be used for parametric amplification. A similar experiment was already successfully shown with a laser amplifier and μJ pulse energies [108]. A further potential application is the temporal compression of such high-power supercontinua. This could allow for the generation of few-cycle laser pulses with multi-Watt average powers with a compact setup.

Appendix A
List of acronyms and abbreviations

A-FPSA	anti-resonant Fabry-Perot saturable absorber
CARS	coherent anti-stokes Raman scattering
CCP	conduction cooled package
COMD	catastrophic optical mirror damage
DM	dichroic mirror
FEA	finite element analysis
FWHM	full width at half maximum
GDD	group delay dispersion
GNLSE	generalized non-linear Schrödinger equation
GTI	Gires-Tournois interferometer mirror (dispersive mirror)
GV	group velocity
GVD	group velocity dispersion
HR	highly reflective mirror
KGW	potassium gadolinium tungstate
KYW	potassium yttrium tungstate
LBO	lithium tri-borate
LOC	large optical cavity concept for broad area diodes
MBE	molecular beam epitaxy
MOCVD	metal-organic chemical vapor deposition
MPC	multi-pass cell
MQW	multiple quantum well
NLSE	non-linear Schrödinger equation
OC	output coupling mirror
PCF	photonic crystal fiber
RMS	root mean square

SC	supercontinuum
SESAM	semi-conductor saturable absorber mirror
SMF28	single mode fiber (Corning)
SPM	self-phase modulation
TEC	thermo-electric cooler
TOD	third-order dispersion
XPM	cross-phase modulation
YAG	yttrium aluminum garnet

Appendix B
List of symbols and constants

$A(t,T)$	($\sqrt{W}$)	normalized field envelope
$A(z,t)$		
$A_{eff,A}$	(m²)	effective mode area on the SESAM
$A_{eff,L}$	(m²)	effective mode area inside the gain medium
$A_{x/y}$	(m⁻¹)	curvature of a parabolic fit
α		taper transition parameter
α	(m⁻¹)	absorption coefficient
α	(K⁻¹)	thermal expansion coefficient
β	(m⁻¹)	fiber propagation constant
β_2	(s²/m)	group velocity dispersion coefficient in fibers
C	(J/kg·K)	specific heat
c	(m/s)	vacuum speed of light
D	(m)	fiber waist diameter
		pump spot diameter
D_2	(s²)	group delay dispersion coefficient (GDD)
δ	(1/W)	modified SPM coefficient
ΔR		saturable reflectivity change (modulation depth)
$\partial n/\partial T$	(K⁻¹)	thermo-optic coefficient
E	(N/m²)	Young's modulus
$E(\mathbf{r},T)$	(V/m)	amplitude of the electric field
E_P	(J)	pulse energy
$E_{sat,A}$	(J)	absorber saturation energy
η		optical-to-optical efficiency
$\mathbf{F}$	(N/m³)	volume force
f	(m)	focal length

Symbol	Unit	Description
$f_{thermal}$	(m)	focal length of the thermal lens
f_{rep}	(s^{-1})	repetition rate
$E_{sat,L}$	(J)	laser saturation energy
$F_{sat,A}$	(J/m^2)	absorber saturation fluence
g		small signal gain
γ	(1/W·m)	SPM coefficient
I	(W/m^2)	laser / pump light intensity
$k(\omega)$	(m^{-1})	wave vector
$k(T)$	(W/mK)	thermal conductivity
k_1	(s/m)	group delay coefficient
k_2	(s^2/m)	group velocity dispersion coefficient
k_3	(s^3/m)	TOD coefficient
L_0	(m)	initial length of the heating zone
		mirror distance in a Herriott-type MPC
l		intensity losses
λ	(m)	wavelength
λ_{ZDW}	(m)	zero-dispersion wavelength
M^2		beam quality parameter ($w_0 \cdot \theta = M^2 \cdot \frac{\lambda}{\pi}$)
μ	(s)	fluorescence lifetime of the upper laser level
N		soliton order
$N_{p/m/g}$		index of refraction along the $p/m/g$-axis
n		index of refraction
n_2	(m^2/W)	nonlinear index
ν	(s^{-1})	light frequency
ν		Poisson's ratio
$\Omega_{(f/g)}$	(s^{-1})	filter/gain bandwidth
ω	(s^{-1})	angular light frequency
ω_0	(s^{-1})	angular center frequency
ω_{dw}	(s^{-1})	angular frequency of the dispersive wave
ω_s	(s^{-1})	angular frequency of the soliton
P	(W)	resonator internal average power
P_{out}	(W)	laser average output power
P_{pump}	(W)	pump power
P_0	(W)	average power
$\hat{P}$	(W)	laser peak power
q		saturable intensity losses

Q	(W/m³)	heat load
R	(m)	mirror radius of curvature
$R(t)$		polarization temporal response function
σ	(N/m²)	stess
σ_{abs}	(m²)	absorption cross section
σ_{em}	(m²)	emission cross section
T	(s)	global time
T	(°C)	temperature
t	(s)	retarded time in respect to the pulse center
T_{mod}	(s)	build-up time of two neighboring modes
T_R	(s)	resonator round-trip time
τ	(s)	width of a sech-pulse ($\tau = 1.76 \cdot \tau_{FWHM}$)
τ_A	(s)	upper-state lifetime of the saturable absorber
τ_{FWHM}	(s)	full width at half maximum pulse duration
τ_L	(s)	lifetime of the upper laser level
τ_{shock}	(s)	self-steepening parameter
θ		far-field divergence angle
v_g	(m/s)	group velocity
$w(z)$	(m)	beam radius in dependence on z
w_0	(m)	beam waist
z	(m)	spatial coordinate in direction of propagation

Appendix C
List of ray-matrices used for resonator calculations and thermal lens retrieval

All simulations of laser resonators in this work as well as the retrieval algorithm of the focal length of the thermal lens were based on the ray-transfer matrix method [48]. The following table provides a list of the employed matrices for different optical elements [48, 109]:

Optical Element	tangential plane	sagittal plane
Linear propagation (length L)	$\begin{pmatrix} 1 & L \\ 0 & 1 \end{pmatrix}$	
Thin lens with focal length f (f>0 convex, perpendicular incidence)	$\begin{pmatrix} 1 & 0 \\ -1/f & 1 \end{pmatrix}$	
Reflection on curved mirror with radius R (R<0: concave), angle of incidence θ	$\begin{pmatrix} 1 & 0 \\ -2/(R \cdot \cos\theta) & 1 \end{pmatrix}$	$\begin{pmatrix} 1 & 0 \\ -2 \cdot \cos\theta/R & 1 \end{pmatrix}$
Flat dielectric interface from medium with n_1 to medium with n_2	$\begin{pmatrix} 1 & 0 \\ 0 & n_1/n_2 \end{pmatrix}$	

Refraction at Brewster angle from medium with n_1 to medium with n_2	$\begin{pmatrix} n_2/n_1 & 0 \\ 0 & n_1^2/n_2^2 \end{pmatrix}$	$\begin{pmatrix} 1 & 0 \\ 1 & n_1/n_2 \end{pmatrix}$
Gradient index lens with $n(x)=n_0-n_0/2\gamma x^2$	$\begin{pmatrix} \cos\gamma z & (n_0\gamma)^{-1}\sin\gamma z \\ -(n_0\gamma)\sin\gamma z & \cos\gamma z \end{pmatrix}$	

Bibliography

[1] J.K. Ranka, R.S. Windeler, and A.J. Stentz, "Visible continuum generation in air-silica microstructure optical fibers with anomalous dispersion at 800 nm," Opt. Lett. **25**, 25–27 (2000).

[2] T.A. Birks, W.J. Wadsworth, and P.St.J. Russel, "Supercontinuum generation in tapered fibers," Opt. Lett. **25**, 1415–1417 (2000).

[3] J.M. Dudley, G. Genty, and S. Coen, "Supercontinuum generation in photonic crystal fiber," Rev. Mod. Phys. **78**, 1135–1184 (2006).

[4] A.V. Husakou and J. Herrmann, "Supercontinuum Generation of Higher-Order Solitons by Fission in Photonic Crystal Fibers," Phys. Rev. Lett. **87**, 203901 (2001).

[5] F. Biancalana, D.V. Skryabin, and A.V. Yulin, "Theory of soliton self-frequency shift compensation by the resonant radiation in photonic crystal fibers," Phys. Rev. E **70**, 70016615 (2004).

[6] J.M. Dudley, L. Provino, N. Grossard, H. Maillotte, R.S. Windeler, B.J. Eggleton, and S. Coen "Supercontinuum generation in air-silica microstructured fibers with nanosecond to femtosecond pulse pumping," J. Opt. Soc. Am. B **19**, 765–771 (2002).

[7] J. Teipel, K. Franke, D. Türke, F. Warken, D Meisner, M. Leuschner, and H. Giessen, "Characteristics of supercontinuum generation in tapered fibers using femtosecond laser pulses," Appl. Phys. B **77**, 245–251 (2003).

[8] D.A. Akimov, A.A. Ivanov, M.V. Alfimov, S.N. Bagayev, T.A. Birks, W.J. Wadsworth, P.St.J. Russell, A.B. Fedotov, V.S. Pivtsov, A.A. Podshivalov, and A.M. Zheltikov, "Two-octave spectral broadening of subnanojoule Cr:forsterite femtosecond laser pulses in tapered fibers," Appl. Phys. B **74**, 307–311 (2002).

[9] J.T. Gopinath, H.M. Shen, H. Sotobayashi, E.P. Ippen, T. Hasegawa, T. Nagashima, and N. Sugimoto, "Highly Nonlinear Bismuth-Oxide Fiber for Supercontinuum Generation and Femtosecond Pulse Compression," J. Lightwave Technol. **23**, 3591–3596 (2005).

[10] N. Nishizawa and J. Takayanagi, "Octave spanning high-quality supercontinuum generation in all-fiber system," J. Opt. Soc. Am. B **24**, 1786–1792 (2007).

[11] J.H. Kim, M.-K. Chen, C.-E. Yang, J. Lee, S. Yin, P. Ruffin, E. Edwards, C. Brantley, and C. Luo, "Broadband IR supercontinuum generation using single crystal sapphire fibers," Opt. Express **16**, 4085–4092 (2008).

[12] B.A. Cumberland, J.C. Travers, S.V. Popov, and J.R. Taylor, "Toward visible cw-pumped supercontinua," Opt. Lett. **33**, 2122–2124 (2008).

[13] J. Teipel, D. Türke, H. Giessen, A. Killi, U. Morgner, M. Lederer, D. Kopf, and M. Kolesik, "Diode-pumped, ultrafast, multi-octave supercontinuum source at repetition rates between 500 kHz and 20 MHz using Yb:glass lasers and tapered fibers," Opt. Express **13**, 1477 (2005).

[14] J. Teipel, D. Türke, H. Giessen, A.Zintl, and B. Braun, "Compact multi-Watt picosecond coherent white light sources using multiple-taper fibers," Opt. Express **13**, 1434 (2005).

[15] A.B. Rulkov, M.Y. Vyatkin, S.V. Popv, J.R. Taylor, and V.P. Gapontsev, "High brightness picosecond all-fiber generation in 525-1800 nm range with picosecond Yb pumping," Opt. Express **13**, 377–381 (2005).

[16] D. Türke, S. Pricking, A. Husakou, J. Teipel, J. Herrmann, and H. Giessen, "Coherence of subsequent supercontinuum pulses in tapered fibers in the femtosecond pumping regime," Opt. Express **15**, 2732 (2007).

[17] E.O. Potma, C.L.Evans, and X.S. Xie, "Heterodyne coherent anti-Stokes Raman scattering (CARS) imaging," Opt. Lett. **31**, 241–243 (2006).

[18] T. Betz, J. Teipel, D. Koch, W. Härtig, J. guck, J. Käs, and H. Giessen, "Excitation beyond the monochromatic laser limit: simultaneous 3-D confocal and multiphoton microscopy with a tapered fiber as white-light laser source," J. Biomed. Opt. **10**, 054009 (2005).

[19] M. Punke, F. Hoos, C. Karnutsch, U. Lemmer, N. Linder, and K. Streubel, "High-repetition-rate white-light pump-probe spectroscopy with a tapered fiber," Opt. Lett. **31**, 1157–1159 (2006).

[20] F. Brunner, T. Südmeyer, E. Innerhofer, F. Morier-Genoud, R. Paschotta, V.E. Kisel, V.G. Shcherbitsky, N.V. Kuleshov, J. Gao, K. Contag, A. Giesen, and U Keller, "240-fs pulses with 22 W average power from a mode-locked thin-disk Yb:KY(WO$_4$)$_2$ laser," Opt. Lett. **27**, 1162–1164 (2002).

[21] G.R. Holtom, "Mode-locked Yb:KGW laser longitudinally pumped by polarization-coupled diode bars," Opt. Lett. **31**, 2719–2721 (2006).

[22] J.E. Hellström, S. Bjurshagen, V. Pasiskevicius, J. Liu, V. Petrov, and U. Griebner, "Efficient Yb:KGW lasers end-pumped by high-power diode bars," Appl. Phys. B **83**, 235–239 (2006).

[23] T. Südmeyer, S.V. Marchese, S. Hashimoto, C.R.E. Baer, G. Gingras, B. Witzel, and U. Keller, "Femtosecond laser oscillators for high-field science," Nature Photonics **2**, 599–604 (2008).

[24] M. Weyers, "GaAs-based high power laser diodes," in "11th European Workshop on MOVPE, Lausanne," pp. 273–278 (2005).

[25] G. Erbert, G. Beister, R. Hülsewede, A. Knauer, W. Pittroff, J. Sebastian, H. Wenzel, M. Weyers, and G. Tränkle, "High-Power Highly Reliable Al-Free 940-nm Diode Lasers," IEEE J. of Sel. Top. Quantum Electron. **7**, 143–148 (2001).

[26] J. Sebastian, G. Beister, F. Bugge, F. Buhrandt, G. Erbert, H.G. Hänsel, R. Hülsewede, A. Knauer, W. Pittroff, R. Staske, M. Schröder, H. Wenzel, M. Weyers, and G. Tränkle, "High-Power 810-nm GaAsP-AlGaAs Diode Lasers With Narrow Beam Divergence," IEEE J. of Sel. Top. Quantum Electron. **7**, 334–339 (2001).

[27] R.L. Sutherland, *Handbook of Nonlinear Optics*, Marcel Dekker, Inc., 2nd edn. (2003).

[28] A.G.P. Agrawal, *Nonlinear fiber optics*, Elsevier Academic Press, 4th edn. (2007).

[29] W.J. Wadsworth, A. Ortigosa-Blanch, J.C. Knight, T.A. Birks, T.-P.M. Man, and P.St.J. Russell, "Supercontinuum generation in photonic crystal fibers and optical fiber tapers: a novel light source," J. Opt. Soc. Am. B **19**, 2148–2155 (2002).

[30] R. Zhang, *Propagation of Ultrashort Light Pulses in Tapered Fibers and Photonic Crystal Fibers*, Ph.D. thesis, Università Bonn, Bonn, Germany (2006), urn:nbn:de:hbz:5N-08619.

[31] D. Türke, *Entstehungsdynamik und Phaseneigenschaften von Weißlicht - Superkontinua aus gezogenen Glasfasern*, Ph.D. thesis, Universität Bonn, Bonn, Germany (2007), urn:nbn:de:hbz:5N-13861.

[32] J. Dudley, X. Gu, L. Xu, M. Kimmel, E. Zeek, P. O'Sheara, R. Trebino, S. Coen, and R.S. Windeler, "Cross-correlation frequency resolved optical gating analysis of broadband continuum generation in photonic crystal fiber: simulations and experiments," Opt. Express **10**, 1215–1221 (2002).

[33] T.A. Birks and Y.W. Li, "The Shape of Fiber Tapers," J. Lightw. Technol. **10**, 432–438 (1992).

[34] F.X. Kärtner, I.D. Jung, and U. Keller, "Soliton Mode-Locking with Saturable Absorbers," IEEE J. Sel. Top. Quantum Electron. **2**, 540–556 (1996).

[35] F.X. Kärtner, J. aus der Au, and U. Keller, "Mode-Locking with Slow and Fast Saturable Absorbers - What's the Difference?" IEEE J. Sel. Top. Quantum Electron. **4**, 159–168 (1998).

[36] U.Keller, D.A.B. Miller, G.D. Boyd, T.H. chiu, J.F. Ferguson, and M.T. Asom, "Solid-state low-loss intracavity saturable absorber for Nd:YLF lasers: an antiresonant semiconductor Fabry-Perot saturable absorber," Opt. Lett. **17**, 505–507 (1992).

[37] F.X. Kärtner, L.R. Brovelli, D. Kopf, M. Kamp, I. Calasso, and U. Keller, "Control of solid state laser dynamics by semiconductor devices," Opt. Eng. **34**, 2024–2036 (1995).

[38] L.R. Brovelli, U. Keller, and T.H. Chiu, "Design and operation of antiresonant Fabry-Perot saturable semiconductor absorbers for mode-locked solid-state lasers," J. Opt. Soc. Am. B **12**, 311–322 (1995).

[39] U.Keller, K.J. Weingarten, F.X. Kärtner, D. Kopf, B. Braun, I.D. Jung, R. Fluck, C. Hönninger, N. Matuschek, and J. aus der Au, "Semiconductor Saturable Absorber Mirrors (SESAM's) for Femtosecond to Nanosecond Pulse Generation in Solid-State Lasers," IEEE J. Sel. Top. Quantum Electron. **2**, 435–451 (1996).

[40] U.Keller, "Recent developments in compact ultrafast lasers," Nature **424**, 831–838 (2003).

[41] H.A. Haus, "Theory of mode locking with a fast saturable absorber," J. Appl. Phys. **46**, 3049–3058 (1975).

[42] C.Hönninger, R. Paschotta, F. Morier-Genoud, M. Moser, and U. Keller, "Q-switching stability limits of continuous-wave passive mode locking," J. Opt. Soc. Am. B **16**, 46–56 (1999).

[43] A.T. Obeidat, W. Knox, and J.B. Khurgin, "Effects of two-photon absorption in saturable Bragg reflectors used in femtosecond solid state lasers," Opt. Express **1**, 68–72 (1997).

[44] W. Richter, BATOP GmbH, Germany, personal communication.

[45] I.D. Jung, F.X. Kärtner, L.R. Brovelli, and U. Keller, "Experimental verification of soliton mode locking using only a slow saturable absorber," Opt. Lett. **20**, 1892–1894 (1995).

[46] M.J. Lederer, B. Luther-Davies, H.H. Tan, and C. Jagadish, "GaAs based anti-resonant Fabry-Perot saturable absorber fabricated by metal organic vapor phase epitaxy and ion implantation," Appl. Phys. Lett. **70**, 3428–3430 (1997).

[47] M.J. Lederer, V. Kolev, B. Luther-Davies, H.T. Tan, and C. Jagadish, "Ion-implanted InGaAs single quantum well semiconductor saturable absorber mirrors for passive mode-locking," J. Phys. D: Appl. Phys. **34**, 2455–2464 (2001).

[48] A.E. Siegman, *Lasers*, University Science Books (1986).

[49] L.M. Osterink and J.D. Foster, "Thermal effects and transverse mode control in a Nd:YAG laser," Appl. Phys. Lett. **12**, 128–131 (1968).

[50] W. Koechner, "Thermal Lensing in a Nd:YAG Laser Rod," Appl. Opt. **9**, 2548–2553 (1970)

[51] M.E. Innocenzi, HT. Yura, C.L. Fincher, and R.A. Fields, "Thermal modeling of continuous-wave end-pumped solid-state lasers," Appl. Phys. Lett. **56**, 1831–1833 (1990).

[52] J.M. Eichenholz and M. Richardson, "Measurements of Thermal Lensing in Cr^{3+}-Doped Colquiriites," IEEE J. Quantum Electron. **34**, 910–919 (1998).

[53] S. Chénais, F. Balembois, F. Druon, G. Lucas-Leclin, and P. Georges, "Thermal Lensing in Diode-Pumped Ytterbium Lasers - Part I: Theoretical Analysis and Wavefront Measurements," IEEE J. Quantum Electron. **40**, 1217–1234 (2004).

[54] W. Koechner, *Solid-State Laser Engineering*, Springer-Verlag Berlin Heidelberg, 5th edn. (1999).

[55] B. Braun, *Compact pulsed diode-pumped solid-state lasers*, Ph.D. thesis, ETH Zürich, Zürich, Switzerland (1996), diss.ETH No.: 11953.

[56] G.Wagner, M. Shiler, and V. Wulfmeyer, "Simulations of thermal lensing of a Ti:Sapphire crystal end-pumped with high average power," Opt. Express **13**, 8045–8055 (2005).

[57] V. Magni, G. Cerullo, and S. De Silvestri, "ABCD matrix analysis of propagation of gaussian beams through Kerr media," Opt. Comm. **96**, 348–355 (1993).

[58] C. Hönninger, F. Morier-Genoud, M. Moser, L.R. Brovelli, and C. Harder, "Efficient and tunable diode-pumped femtosecond Yb:glass lasers," Opt. Lett. **23**(2), 126–128 (1998).

[59] C. Hönninger, R. Paschotta, M. Graf, F. Morier-Genoud, G. Zhang, M. Moser, S. Biswal, J. Nees, A. Braun, G.A. Mourou, I. Johannsen, A. Giesen, W. Seeber, and U. Keller, "Ultrafast ytterbium-doped bulk lasers and laser amplifiers," Appl. Phys. B **69**, 3–17 (1999).

[60] D.J. Richardson, R. Paschotta, and D.C. Hanna, "Stretched pulse Yb^{3+}:silica fiber laser," Opt. Lett. **22**, 316–318 (1997).

[61] J.R. Buckley, F.W. Wise, F.Ö. Ilday, and T. Sosnowski, "Femtosecond fiber lasers with pulse energies above 10 nJ," Opt. Lett. **30**, 1888–1890 (2005).

[62] D. Türke, W. Wohlleben, J. Teipel, M. Motzkus, B Kibler, J. Dudley, and H. Giessen, "Chirp-controlled soliton fission in tapered optical fibers," Appl. Phys. B **83**, 37–42 (2006).

[63] C. Florea and K.A. Winick, "Ytterbium-Doped Glass Waveguide Laser Fabricated by Ion Exchange," J. Lightwave Technol. **17**, 1593–1601 (1999).

[64] S. Jiang, M.J. Myers, D.L. Rhonehouse, U. Griebner, R. Koch, and H. Schonnagel, "Ytterbium doped phosphate laser glass," in Richard Scheps, ed., "Solid State Lasers VI (Proceedings Volume)," pp. 10–15 (1997).

[65] Kigre Inc., "QX laser glasses," datasheet.

[66] S. Liu and A. Lu, "Physical and Spectroscopic Properties of Yb^{3+}-Doped Fluorophosphate Laser Glasses," Laser Chemistry **2008**, 1–6 (2008).

[67] Kigre Inc., personal communication.

[68] R. Koch, W.A. Clarkson, D.C. Hanna, S. Jiang, M.J. Myers, D. Rhonehouse, S.J. Hamlin, U. Griebner, and H. Schönnagel, "Efficient room temperature cw Yb:glass laser pumped by a 946 nm Nd:YAG laser," Opt. Comm. **134**, 175–178 (1997).

[69] A.E. Siegman, "Analysis of laser beam quality degradation caused by quartic phase aberrations," Appl. Opt. **32**(30), 5893–5901 (1993).

[70] D.R. Herriott and H.J. Schulte, "Folded Optical Delay Lines," Appl. Opt. **4**, 883–889 (1965).

[71] A. Sennaroglu and J.G. Fujimoto, "Design criteria for Herriott-type multi-pas cavities for ultrashort pulse lasers," Opt. Express **11**, 1106–1113 (2003).

[72] K. Naganuma, G. Lenz, and E.P. Ippen, "Variable Bandwidth Birefringent Filter for Tunable Femtosecond Lasers," IEEE J. Quantum Electron. **28**, 2141–2150 (1992).

[73] J.A. Berger, M.J.Greco, and W.A. Schroeder, "High-Power, Femtosecond, Thermal-Lens-Shaped Yb:KGW Oscillator," Opt. Express **16**, 8629 (2008).

[74] J. Neuhaus, D. Bauer, J. Zhang, A. Killi, J. Kleinbauer, M. Kumkar, S. Weiler, M. Guina, D.H. Sutter, and T. Dekorsy, "Subpicosecond thin-disk laser oscillator with pulse energies of up to 25.9 microjoules by use of an active multipass geometry," Opt.Express **16**, 20530–20539 (2008).

[75] T. Kawashima, T. Kurita, O. Matsumoto, T. Ikegawa, T. Sekine, M. Miyamoto, K. Iyama, H. Kan, and Y. Tsuchiya, "20-J Diode-Pumped Zig-Zag Slab Laser with 2-GW Peak Power and 200-W Average Power," in "ASSP 2005," vol. TuB44 (2005).

[76] X. Mateos, V. Petrov, M. Aguiló, R. Solé, J. Gavaldà, J. Massons, F. Díaz, and U. Griebner, "Continuous-Wave Laser Oscillation of Yb^{3+} in Monoclinic $KLu(WO_4)_2$," IEEE J. of Quantum Electron. **40**, 1056–1059 (2004).

[77] U. Griebner, J. Liu, S. Rivier, A. Aznar, R. Grunwald, R.M. Solé, M. Aguiló, F. Díaz, and V. Petrov, "Laser Operation of Expitaxially Grown $Yb:KLu(WO_4)_2$-$KLu(WO_4)_2$ Composites With Monoclinic Crystalline Structure," IEEE J. of Quantum Electron. **41**, 408–414 (2005).

[78] S. Rivier, X. Mateos, O. Silvestre, V. Petrov, U. Griebner, M.C. Pujol, M. Aguiló, F. Díaz, S. Vernay, and D. Rytz, "Thin-disk $Yb:KLu(WO_4)_2$ laser with single-pass pumping," Opt. Lett. **33**, 735–737 (2008).

[79] EKSPLA, "Yb:KGW and Yb:KYW crystals," datasheet, http://www.ekspla.com/en/main/products/?PID=498 (2008).

[80] I.V. Mochalov, "Nonlinear optics of the potassium gadolinium tungstate laser crystal $Nd^{3+}:KGd(WO_4)_2$," J. Opt. Technol. **62**, 746–756 (1995).

[81] C. Hönninger, R. Paschotta, M. Graf, F. Morier-Genoud, G. Zhang, M. Moser, S. Biswal, J. Nees, A. Braun, G.A. Mourou, I. Johannsen, A. Giessen, W. Seeber, and U. Keller, "Ultrafast ytterbium-doped bulk lasers and laser amplifiers," Appl. Phys. B **69**, 3–17 (1999).

[82] H. Liu, J. Nees, and G. Mourou, "Diode-pumped Kerr-lens mode-locked $Yb:KY(WO_4)_2$ laser," Opt. Lett. **26**, 1723–1725 (2001).

[83] A. Brenier and G. Boulon, "New criteria to choose the best Yb^{3+}-doped laser crystal," Europhys. Lett. **55**, 647–652 (2001).

[84] M.C. Pujol, M. Rico, C. Zaldo, R. Solé, V. Nikolov, X. Solans, M. Aguiló, and F. Díaz, "Crystalline structure and optical spectroscopy of Er^{3+}-doped $KGd(WO_4)_2$ single crystals," Appl. Phys. B **68**, 187–197 (1999).

[85] F. Brunner, G.J. Spühler, J. Aus der Au, L. Krainer, F. Morier-Genoud, R. Paschotta, N. Lichtenstein, S. Weiss, C. Harder, A.A. Lagatsky,

A. Abdolvand, N.V. Kuleshov, and U. Keller, "Diode-pumped femtosecond Yb:KGd(WO$_4$)$_2$ laser with 1.1-W average power," Opt. Lett. **25**, 1119–1121 (2000).

[86] M.C. Pujol, R. Solé, J. Massons, J. Gavaldà, X. Solans, C. Zaldo, F. Díaz, and M. Aguiló, "Structural study of monoclinic KGd(WO$_4$)$_2$ and effects of lanthanide substitution," J. Appl. Cryst. **34**, 1–6 (2001).

[87] A. Killi, *Cavity dumping in solitary mode-locked femtosecond laser oscillators*, Ph.D. thesis, Universität Heidelberg, Heidelberg, Germany (2005), urn:nbn:de:bsz:16-opus-56402.

[88] A.A. Lagatsky, A. Abdolvand, and N.V. Kuleshov, "Passive Q switching and self-frequency Raman conversion in a diode-pumped Yb:KGd(WO$_4$)$_2$ laser," Opt. Lett. **25**, 616–618 (2000).

[89] D. Schröder, J. Meusel, P. Hennig, D. Lorenzen, M. Schröder, R. Hülsewede, and J. Sebastian, "Increased power of broad-area lasers (808nm/980nm) and applicability to 10mm-bars with up to 1000 Watt QCW," in M.S. Zediker, ed., "High-Power Diode Laser Technology and Applications V," vol. 6456, Proc. of SPIE (2007).

[90] D. Kopf, K.J. Weingarten, G. Zhang, M. Moser, M.A. Emanuel, R.J. Beach, J.A. Skidmore, and U. Keller, "High-average-power diode-pumped femtosecond Cr:LiSAF lasers," Appl. Phys. B **65**, 235–243 (1997).

[91] J. Aus der Au, S.F. Schaer, R. Paschotta, C. Hönninger, and U. Keller, "High-power diode-pumped passively mode-locked Yb:YAG lasers," Opt. Lett. **24**, 1281–1283 (1999).

[92] C. Jacinto, T. Catunda, D. Jaque, L.E. Bausá, and J. García-Solé, "Thermal lens and heat generation of Nd:YAG lasers operating at 1.064 and 1.34 μm," Opt. Express **16**, 6317–6323 (2008).

[93] S.M. Mian, S.B. McGee, and N. Melikechi, "Experimental and theoretical investigation of thermal lensing effects in mode-locked femtosecond Z-scan experiments," Opt. Comm. **207**, 339–345 (2002).

[94] G.L.Bourdet, I. Hassiaoui, J.F. Monjardin, H. Baker, N. Michel, and M. Krakowski, "High-power, low-divergence, linear array of quasi-diffraction-limited beams supplied by tapered diodes," Appl. Opt. **46**, 6297–6301 (2007).

[95] V.V. Filippov, N.V. Kuleshov, and I.T. Bodnar, "Negative thermo-optical coefficients and athermal directions in monoclinic KGd(WO$_4$)$_2$ and KY(WO$_4$)$_2$ laser host crystals in the visible region," Appl. Phys. B **87**, 611–614 (2007).

[96] S. Biswal, S. Shawn, P. O'Connor, and S.R. Bowman, "Thermo-optical parameters measured in potassium-gadolinium-tungstate," in "CLEO 2004," vol. CThT62 (2004).

[97] K.V. Yumashev, V.G. Savitski, N.V. Kuleshov, A.A. Pavlyuk, D.D. Molotkov, and A.L. Protasenya, "Laser performance of N_g-cut flash-lamp pumped Nd:KGW at high repetition rates," Appl. Phys. B **89**, 39–43 (2007).

[98] S. Biswal, S.P. O'Connor, and S.R. Bowman, "Thermo-optical parameters measured in ytterbium-doped potassium gadolinium tungstate," Appl. Opt. **44**, 3093–3097 (2005).

[99] H. Ibach and H. Lüth, *Solid-State Physics*, Springer-Verlag Berlin Heidelberg, 3rd edn. (2003).

[100] R.L. Aggarwal, D.J. Ripin, J.R. Ochoa, and T.Y. Fan, "Measurement of thermo-optic properties of Y$_3$Al$_5$O$_1$2,Lu$_3$Al$_5$O$_1$2,YAlO$_3$,LiYF$_4$,LiLuF$_4$,ByY$_2$F$_8$,KGd(WO$_4$)$_2$, and KY(WO$_4$)$_2$ laser crystals in the 80-300 K temperature range," J. Appl. Phys. **98**, 103514 (2005).

[101] C.E. Mungan, "Thermodynamics of radiation-balanced lasing," J. Opt. Soc. Am. B **20**, 1075–1082 (2003).

[102] M. Okamoto and H. Sasada, "Generation of Optical Vortices by Converting Elegant Hermite Gaussian Beams," Jpn. J. Appl. Phys. **44**, 1743–1747 (2005).

[103] J.E. Hellström, S. Bjurshagen, and V. Pasiskevicius, "Laser performance and thermal lensing in high-power diode-pumped Yb:KGW with athermal orientation," Appl. Phys. B **83**, 55–59 (2006).

[104] H.A. Haus and E.P. Ippen, "Self-starting of passively mode-locked lasers," Opt. Lett. **16**, 1331–1333 (1991).

[105] K.L. Corwin, N.R. Newbury, J.M. Dudley, S. Coen, S.A. Diddams, B.R. Washburn, K. Weber, and R.S. Windeler, "Fundamental amplitude noise limitations

to supercontinuum spectra generated in a microstructured fiber," Appl. Phys. B **77**, 269 (2003).

[106] K.L. Corwin, N.R. Newbury, J.M. Dudley, S. Coen, S.A. Diddams, B.R. K. Weber, and R.S. Windeler, "Fundamental Noise Limitations to Supercontinuum Generation in Microstructure Fiber," Phys. Rev. Lett. **90**, 113904 (2003).

[107] D. Mogilevtsev, J. Broeng, S.E. Barkou, and A. Bjarklev, "Design of polarization-preserving photonic crystal fibres with elliptical pores," J. Opt. A: Pure Appl. Opt. **3**, S141–S143 (2001).

[108] J. Limpert, C. Aguergaray, S. Montant, I. Manek-Hönninger, S. Petit, D. Descamps, E. Cormier, and F. Salin, "Ultra-broad bandwidth parametric amplification at degeneracy," Opt. Express **13**, 7386–7392 (2005).

[109] H.T. Yura and S.G. Hanson, "Optical beam wave propagation through complex optical systems," J. Opt. Soc. Am. A **4**, 931–1948 (1987).

Acknowledgements

Finally, I would like to express my gratitude to all people who contributed to the success of this work. Many thanks go to...

- Prof. Dr. Harald Giessen for supervising me and supporting me for many years, for the confidence he put in me, for the freedom he gave me to also pursue my own ideas, and his never ending enthusiasm and passion for physics.

- Prof. Dr. Bernd Braun for his invaluable support and advice, for proof-reading my writings so many times at any possible daytime, and for the many fruitful discussions we had.

- Prof. Dr. Peter Michler for kindly agreeing to be my second reviewer.

- Prof. Dr. Hans Peter Büchler for kindly agreeing to be the head of the examination board.

- Dr. Todd Meyrath for all the things he taught me about electronics, for correcting my English so many times, and for the great time and many long movie nights in Stuttgart.

- Sebastian Pricking for many helpful discussions, the great hiking tours, and for being my office colleague for a long time.

- Gabi Feurle and Dr. Christine von Rekowski for their support in all administrative matters.

- Herr Kamella and all the people from the mechanical workshop for their support.

- My colleagues from the 4th physics institute (in alphabetical order): Ralf Ameling, Christina Bauer, Tilman Benkert, Daniel Dregely, Bettina Frank, Liwei Fu, Michael Geiselmann, Hedi Gräbeldinger, Cornelius Grossmann, Hongcang Guo, Robin Hegenbarth, Georg Kobiela, Friederike Kromer, Daniel Kunert,

Lutz Langguth, Sai Li, Dr. Markus Lippitz, Na Liu, Thomas Luckert, Patrick Mai, Martin Mesch, Bernd Metzger, Todd Meyrath, Regina Orzekowsky, Markus Pfeiffer, Tim Preukschat, Sebastian Pricking, Dr. Heinz Schweizer, Andreas Seidel, Richard Taubert, Monika Ubl, Tobias Utikal, Marius Vieweg, Anja Wacker, Thomas Weiss, Thomas Zentgraf.

- My former colleagues in Bonn: Dietmar Nau, Anja Schönhardt, Jörn Teipel, Diana Türke, Rui Zhang.

- Meinen Eltern, denen ich soviel zu verdanken habe und die mich während des ganzen langen Weges unterstützt haben.

- Und zu guter Letzt meiner Freundin Sina dafür, dass sie die ganze Zeit an meiner Seite gestanden, mich immer liebevoll unterstützt und auch in hektischen Zeiten ertragen hat. Ohne Dich wäre die Zeit niemals so schön gewesen!

Die VDM Verlagsservicegesellschaft sucht für wissenschaftliche Verlage abgeschlossene und herausragende

Dissertationen, Habilitationen, Diplomarbeiten, Master Theses, Magisterarbeiten usw.

für die kostenlose Publikation als Fachbuch.

Sie verfügen über eine Arbeit, die hohen inhaltlichen und formalen Ansprüchen genügt, und haben Interesse an einer honorarvergüteten Publikation?

Dann senden Sie bitte erste Informationen über sich und Ihre Arbeit per Email an *info@vdm-vsg.de*.

Sie erhalten kurzfristig unser Feedback!

VDM Verlagsservicegesellschaft mbH
Dudweiler Landstr. 99
D - 66123 Saarbrücken
Telefon +49 681 3720 174
Fax   +49 681 3720 1749

www.vdm-vsg.de

Die VDM Verlagsservicegesellschaft mbH vertritt

Printed by Books on Demand GmbH, Norderstedt / Germany